CHEMISTRY RESEARCH AND APPLICATIONS

BERYLLIUM

PHYSICOCHEMICAL PROPERTIES, APPLICATIONS AND SAFETY CONCERNS

CHEMISTRY RESEARCH AND APPLICATIONS

Additional books in this series can be found on Nova's website under the Series tab.

Additional e-books in this series can be found on Nova's website under the e-book tab.

CHEMISTRY RESEARCH AND APPLICATIONS

BERYLLIUM

PHYSICOCHEMICAL PROPERTIES, APPLICATIONS AND SAFETY CONCERNS

PAULEEN DYER
EDITOR

New York

For permission to use material from this book please contact us:
Telephone 631-231-7269; Fax 631-231-8175
Web Site: http://www.novapublishers.com

Library of Congress Cataloging-in-Publication Data

Beryllium (Nova Science Publishers)
Beryllium : physicochemical properties, applications and safety concerns / [edited by] Pauleen Dyer.
pages cm -- (Chemistry research and applications)
Includes bibliographical references and index.
ISBN: 978-1-63321-590-0 (softcover)
1. Beryllium. I. Dyer, Pauleen, editor. II. Title.
QD181.B4B455 2014
546'.391--dc23
2014027185

Published by Nova Science Publishers, Inc. † New York

CONTENTS

PREFACE

Beryllium (Be) is a lightweight metal with unique chemical and physical properties. It is stiff, resistant to corrosion, a good conductor of electricity and heat, and dimensionally stable during extreme temperature changes. These properties have made it desirable for many commercial and defense applications, such as aircraft and satellite structures, high-speed electronic circuitry, nuclear applications, and precision instruments. It has also been used in dental appliances, golf clubs, non-sparking tools, and wheelchairs. This book discusses the physicochemical properties, applications and safety concerns of beryllium.

Chapter 1 - The development and adoption of the beryllium lymphocyte proliferation test (BeLPT) during the 1980s made it possible to identify asymptomatic individuals who were sensitized to beryllium and either had chronic beryllium disease (CBD) or were at risk for developing it. The BeLPT has been used extensively for the surveillance of beryllium workers. However, inconsistency in BeLPT test results has led some physicians to require more than one abnormal test result before categorizing an individual as "sensitized."

While planning to test the community surrounding a large beryllium producer, the Agency for Toxic Substances and Disease Registry convened an expert panel and asked individual members to provide opinions on how best to apply and interpret BeLPT testing. There were three different opinions for the appropriate minimum criteria to use for confirming beryllium sensitization:

1) one abnormal BeLPT,
2) one abnormal BeLPT and one borderline BeLPT, and
3) two abnormal BeLPTs.

Since that time, the authors have developed a testing plan for each of these three minimum criteria. Each plan contains the relevant minimum criteria, a flow diagram with the corresponding decision logic, and a table to quantify overall performance. Finally, the authors calculated the overall sensitivity and specificity for each plan, as well as the plan's positive predictive values across a range of assumed population prevalences.

These epidemiologic parameters are affected by requiring confirmation beyond one abnormal BeLPT, either as a borderline or as another abnormal. Confirmation raises specificity and lowers sensitivity, while providing higher positive predictive values at any given prevalence. The purposes for testing may affect the relative importance of these various parameters.

This chapter predicts and compares the performance of three plans for testing exposed groups that are asymptomatic. When other factors are present, such as respiratory signs or symptoms consistent with CBD, clinical considerations may dictate a different approach.

Chapter 2 - Chronic beryllium disease (CBD) can occur by exposure to beryllium, a light earth alkaline metal. Beryllium and its compounds are used in the electronic-, computer-, atomic-, and ceramics industries. Beryllium is also found in fossil fuels and can be released through combustion. The uptake of beryllium and its compounds occur mainly by inhalation. Acute beryllium disease (ABD) is a toxic chemical reaction to beryllium and its compounds and mainly effects the respiratory tract. CBD is a specific cell-mediated delayed immune response on beryllium. The susceptibility of CBD is associated with the HLA-DPB1 gene possessing glutamic acid at 69th position (E69) of the ß-chain. Pathophysiologically non-necrotizing granulomas are detectable most commonly in lung and skin. The main symptoms include dry cough, fatigue, weight loss and increasing shortness of breath. CBD can be mistaken for sarcoidosis due to the similar symptoms and the formation of granulomas. However, the beryllium lymphocyte proliferation test in CBD is positive in contrast to sarcoidosis. Therapeutic steroids are used in addition to avoidance of beryllium exposure. Even if there is no further exposure, CBD can also progress over decades and end in pulmonary fibrosis. Beryllium and its compounds are considered to be carcinogenic, because in patients suffering from ABD or CBD and in individuals exposed to beryllium an increased rate of lung cancer was observed.

Chapter 3 - The application of cosmogenic nuclide beryllium-10 (^{10}Be) in marginal sea study has been discussed. Dissolved ^{10}Be concentration profiles in seawaters of the East China Sea and the Kuroshio Current have been investigated. The results show that ^{10}Be concentrations in this area are mainly controlled by surface biological productivity, particle remineralization, and the degree of mixing with the Yangtze River and the Kuroshio waters. Generally the ^{10}Be water depth profiles can be divided into three layers: the surface mixing layer, the particulate ^{10}Be regeneration layer and the bottom layer. Vertical distributions of ^{10}Be show that ^{10}Be is enriched in the bottom waters near the Yangtze River estuary and the central continental shelf. Box model results indicate that ^{10}Be input from the Kuroshio Current is more important than the Yangtze River input and atmospheric precipitation. About 81% of the ^{10}Be input to the East China Sea is scavenged into the sediments and 19% of the ^{10}Be flows out of the East China Sea by currents and water exchange. The ^{10}Be sedimentation flux in the East China Sea is nearly five times higher than the average global ^{10}Be production rate. Therefore the East China Sea may be an important sink for ^{10}Be. ^{10}Be records in two sediment cores from the Okinawa Trough during the last glacial period and the Holocene are discussed. The average ^{10}Be sedimentation flux during the last glacial period was more than three times higher than that at present and also higher than the ^{10}Be accumulation fluxes in the Pacific open ocean during the glacial period, indicating the conveyor rule of the Kuroshio Current and a boundary scavenging effect for ^{10}Be. The high ^{10}Be concentrations and fluxes in the Okinawa Trough during the glacial period proved that on millennium scale the Kuroshio Current still flowed in the Okinawa Trough area with significantly high intensity. The ^{10}Be minimum during the Younger Dryas cold event may indicate that the Kuroshio had been once greatly weakened or even terminated in the Okinawa Trough area which could be related to the response of the Pacific Ocean to the Younger Dryas.

Chapter 4 - Beryllium-7 (^{7}Be) is a relatively short-lived radionuclide (half-life 53.3 days) which decays by electron capture either directly to the ground state of ^{7}Li (89.56%) or to an excited state of ^{7}Li (10.44%), which decays to the ground state of ^{7}Li via gamma-ray emission at 477.6 keV. This allows us to easily quantify it by using gamma-ray spectrometers. Beryllium-7 has a cosmogenic origin and is produced in the upper atmosphere and lower stratosphere by high-energy spallation interactions of nitrogen and oxygen. It continuously enters to marine and terrestrial ecosystems via wet (over 90%) and dry (3 to 10%) deposition. Several factors can affect this input, such as production rate (which varies with latitude, altitude, and solar activity), stratosphere–troposphere mixing, circulation and advection processes within the troposphere and efficiency with which it is removed from the troposphere. After deposition, ^{7}Be will tend to associate with particulate material (particle-reactive element). Its relatively short half-life, reactivity, and continuous and definable production rates make ^{7}Be a potentially powerful tool for the study and description of several environmental processes such as soil redistribution, sediment sources assessment, concentration in air, air mass transport, study of metal scavenging and others. In order to use ^{7}Be as an environmental tracer, the knowledge of its input from the atmosphere and its variability are needed. However, when its input is evaluated, divergent information may be obtained. For different regions, dissimilar environmental conditions and seasons of the year, ^{7}Be rain water content shows a high variability, and the cause of these changes could be difficult to understand or explain. A high ^{7}Be content has been reported for some environments for precipitations of a few millimeters and low ^{7}Be contents for precipitations occurring after other precipitation event. These results have been explained by the atmospheric washing phenomenon and a reload rate can be estimated. Moreover, effects of rainfall rate on rain ^{7}Be content have been reported with divergent results. Despite these, there is agreement that wet deposition on the ground can be estimated from the rainfall volume. This chapter summarizes the results obtained in evaluating the ^{7}Be content in rainfalls for a semiarid environment characterized by a seasonal precipitation regime. For entire rain events, the effect of precipitation variables on ^{7}Be content in rain water is evaluated and contrasted with other regions. For single rain events the changes of ^{7}Be content and the effect of rainfall intensity is evaluated for each millimeter of rain fallen.

In: Beryllium
Editor: Pauleen Dyer
ISBN: 978-1-63321-590-0

Chapter 1

PREDICTING PERFORMANCE FOR BeLPT TESTING: A REVIEW

Dan Middleton**, M.D., MPH and Michael Lewin*†, MS
Agency for Toxic Substances and Disease Registry

ABSTRACT

The development and adoption of the beryllium lymphocyte proliferation test (BeLPT) during the 1980s made it possible to identify asymptomatic individuals who were sensitized to beryllium and either had chronic beryllium disease (CBD) or were at risk for developing it. The BeLPT has been used extensively for the surveillance of beryllium workers. However, inconsistency in BeLPT test results has led some physicians to require more than one abnormal test result before categorizing an individual as "sensitized."

While planning to test the community surrounding a large beryllium producer, the Agency for Toxic Substances and Disease Registry convened an expert panel and asked individual members to provide opinions on how best to apply and interpret BeLPT testing. There were three different opinions for the appropriate minimum criteria to use for confirming beryllium sensitization:

* Email: dcm2@cdc.gov.
† Email: mdl0@cdc.gov.

4) one abnormal BeLPT,
5) one abnormal BeLPT and one borderline BeLPT, and
6) two abnormal BeLPTs.

Since that time, we have developed a testing plan for each of these three minimum criteria. Each plan contains the relevant minimum criteria, a flow diagram with the corresponding decision logic, and a table to quantify overall performance. Finally, we calculated the overall sensitivity and specificity for each plan, as well as the plan's positive predictive values across a range of assumed population prevalences.

These epidemiologic parameters are affected by requiring confirmation beyond one abnormal BeLPT, either as a borderline or as another abnormal. Confirmation raises specificity and lowers sensitivity, while providing higher positive predictive values at any given prevalence. The purposes for testing may affect the relative importance of these various parameters.

This chapter predicts and compares the performance of three plans for testing exposed groups that are asymptomatic. When other factors are present, such as respiratory signs or symptoms consistent with CBD, clinical considerations may dictate a different approach.

INTRODUCTION

Beryllium (Be) is a lightweight metal with unique chemical and physical properties. It is stiff, resistant to corrosion, a good conductor of electricity and heat, and dimensionally stable during extreme temperature changes. These properties have made it desirable for many commercial and defense applications, such as aircraft and satellite structures, high-speed electronic circuitry, nuclear applications, and precision instruments. It has also been used in dental appliances, golf clubs, non-sparking tools, and wheelchairs (ATSDR 2002).

Chronic beryllium disease (CBD) was first observed in the United States in 1946 among workers manufacturing fluorescent lamps (Hardy and Tabershaw 1946). CBD is a granulomatous lung disease that results from inhaling airborne beryllium particles; it typically develops over months or years from a hypersensitivity response that can occur even at relatively low exposure levels. The latency period can be long, and CBD is relatively difficult to diagnose (Newman et al. 2005). The BeLPT plays an important role in the differential diagnosis of granulomatous lung disease (Middleton 1998), but in this chapter we focus on exposed groups that are asymptomatic.

Uses of the BeLPT

Introduction of the beryllium lymphocyte proliferation test (BeLPT) during the 1980s made it possible to identify individuals who were sensitized to beryllium and either had or were at risk for developing CBD (Kreiss et al. 1989). The BeLPT has been used extensively for the surveillance of beryllium workers. In recent years it has also been used to evaluate the effect of interventions designed to lower exposures in beryllium facilities, including a ceramic oxide plant (Cummings et al. 2007), a primary beryllium materials production plant (Bailey et al. 2010), and a copper-beryllium facility (Thomas et al. 2009).

Beryllium exposure is not limited to workers who produce beryllium metal and its alloys. The National Institute for Occupational Safety and Health (NIOSH), Centers for Disease Control and Prevention (CDC), estimated that 26,400 to 134,000 workers may be exposed to more than 0.1 $\mu g/m^3$ of beryllium in air (Henneberger et al. 2004). Beryllium sensitization may occur at even lower levels, and the number of workers (and others) at risk may be much higher than estimated. Unrecognized exposures outside the workplace have also occurred from worker take-home and from industrial releases into the environment (Maier et al. 2008).

Once sensitivity to beryllium is established with the BeLPT, the sensitized individual is usually evaluated clinically for granulomatous lung disease. The medical tests included in the workup can be varied on a case by case basis, but the evaluation usually includes a chest radiograph, pulmonary function testing, bronchoscopy with lavage, and a lung biopsy. Typically, another BeLPT is performed on T-lymphocytes collected from the lungs during the bronchoscopy (Maier 2001).

Genetic Susceptibility

There is evidence that the genetic susceptibility of exposed individuals is an important factor in developing CBD, although exposure to beryllium is a necessary determinant. CBD is associated with a major histocompatibility complex (MHC) class II marker (HLA-DPB1Glu69) (Richeldi et al. 1997). While this marker is still considered a very important risk factor, it is not the sole determinant of beryllium sensitization and CBD. Exposure level continues to be a significant factor, and progress has continued in clarifying the relationship between various other genetic markers and the risk for beryllium sensitization and disease (Rossman et al. 2002; Silveira et al. 2012).

The BeLPT

Performance of the BeLPT is described in detail in numerous publications (Kreiss et al. 1989; Mroz et al. 1991; Maier 2001). Briefly, to perform a single BeLPT, an individual's T-lymphocytes are incubated in three concentrations of beryllium sulfate over two time periods (six incubations). If the individual's T-lymphocytes are sensitized, incubations with beryllium sulfate will take up more tritiated thymidine (thymidine containing the radioisotope tritium) than will incubations without beryllium sulfate. The radioactivity levels of the incubations are compared, and results are interpreted as follows:

1. normal test result (NL) – 0 of 6 incubations with beryllium sulfate are elevated,
2. borderline test result (BL) – 1 of 6 incubations with beryllium sulfate is elevated, and
3. abnormal test result (AB) – 2 or more incubations with beryllium sulfate are elevated.

An abnormal result can occur when the individual is truly sensitized (true positive) or as a test error (false positive). Also, a normal result can occur when the individual is not truly sensitized (true negative) or as a test error (false negative).

Generally speaking, an abnormal BeLPT result identifies an individual who is at risk for CBD. However, Deubner et al. (2001) found substantial intra- and inter-laboratory disagreement among the laboratories they evaluated. Concern about this lack of consistency has led many practitioners to require multiple tests and to specify various combinations of BeLPT results to confirm immunologic sensitivity to beryllium.

There is still no consensus on the combination of BeLPT results that provides the "best" criteria for determining sensitization. To address this issue, the Agency for Toxic Substances and Disease Registry convened an expert panel in Ottawa County, Ohio, in 2006. This panel included physicians knowledgeable about beryllium from environmental groups, industry, universities, and government. When asked about the appropriate criteria for confirming beryllium sensitization, the panelists were divided among the three criteria below (Middleton et al. 2008):

a) one abnormal BeLPT,
b) one abnormal BeLPT and one borderline BeLPT, and
c) two abnormal BeLPTs.

In this chapter we focus on three plans designed to implement testing in exposed groups that are asymptomatic.

Performance Characteristics of the BeLPT

The performance of a single test or a series of tests (testing program) is best described by quantitative epidemiologic parameters, including sensitivity, specificity, and positive predictive value (PPV). In this chapter we will use the following symbols:

1. P_{AB} = probability of an abnormal result,
2. P_{BL} = probability of a borderline result, and
3. P_{NL} = probability of a normal result.

Note that the probabilities for test outcomes (P_{AB}, P_{BL}, and P_{NL}) will differ for the two categories of true immunologic sensitization (truly sensitized / not truly sensitized). While there is no gold standard, Stange et al. (2004) proposed the following (in general terms) as a working standard for serial testing results:

1. if an AB was not confirmed by a second AB, the AB was a false positive;
2. if an AB was confirmed by a second AB, the participant was sensitized;
3. if an AB was confirmed by a second AB, a NL within 2 years was a false negative.

The implicit assumption was that serial testing will confirm a truly positive result (2 AB), while additional testing will usually not confirm a *false* positive result (1 AB). This logic was applied to a database containing 19,396 serial BeLPT results collected over 10 years for 7820 current and former employees who were potentially exposed to beryllium. After repeating borderline results, the researchers estimated that the probability of an abnormal BeLPT (P_{AB}) for someone truly sensitized to Be (test sensitivity) was 0.683 and the probability of a normal result (P_{NL}) for someone not truly sensitized to Be (test specificity) was 0.969.

Middleton et al. (2006, 2008) adjusted these probabilities to include borderline test results. For individuals truly sensitized to beryllium, the test outcome probabilities are $P_{AB} = 0.5970$; $P_{NL} = 0.2770$; and $P_{BL} = 0.1260$. The test outcome probabilities for individuals not truly sensitized to beryllium are $P_{AB} = 0.0109$; $P_{NL} = 0.9733$; and $P_{BL} = 0.0158$. These probabilities allow us to quantify the three testing plans and compare these plans quantitatively using epidemiologic parameters.

TESTING PLANS

In this chapter there are three well-defined plans for testing with the BeLPT; each one is based on the associated minimal criteria as follows: Plan 1 (1 AB), Plan 2 (1 AB and 1 BL), and Plan 3 (2 AB). Each testing plan also includes a figure illustrating the flow of testing and a table to quantify the plan's overall performance.

While we demonstrated in a previous publication that it may not be ideal (Middleton et al. 2006), most testing begins with a single BeLPT. After the first BeLPT, testing continues following a predefined algorithm specific to the testing plan selected (Figures 1, 2, and 3). We note that Plan 1 and Plan 3 both treat borderline results as placeholders to be repeated until an abnormal or a normal result occurs.

Plan 1. One Abnormal BeLPT

Figure 1 illustrates the flow of testing when one abnormal BeLPT is considered sufficient evidence for sensitization. Note that if the initial test is abnormal, the criteria for sensitization are met (one abnormal). If a normal result is obtained, testing ends for this round and the individual is considered currently negative for sensitization.

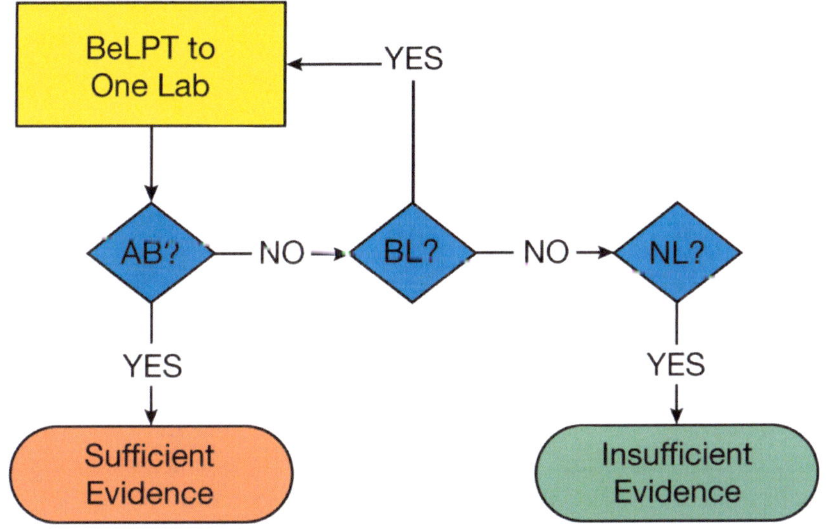

Figure 1. One abnormal BeLPT provides sufficient evidence for beryllium sensitization (Plan 1) (Middleton and Kowalski 2010).

Table 1 contains the probability calculations for Plan 1. Consider the case of individuals who are *truly sensitized.* The sensitivity of the plan overall (P_{Total}) is the sum of the probabilities for the following: an AB as the first result (P_{AB}), a BL followed by an AB ($P_{BL\ then\ AB}$), and 2 BL followed by an AB ($P_{2BL\ then\ AB}$).That is,

a) $P_{AB} = 0.5970$;
b) $P_{BL\ then\ AB} = P_{BL} \times P_{AB} = 0.1260 \times 0.5970 = 0.0752$;
c) $P_{2BL\ then\ AB} = P_{BL}^2 \times P_{AB} = 0.1260^2 \times 0.5970 = 0.0095$; and,
d) $P_{Total} = P_{AB} + P_{BL\ then\ AB} + P_{2BL\ then\ AB} = 0.6817$.

The probability that *truly sensitized* individuals are correctly identified is approximately 0.6817, or 68.2%. While there are many more potential combinations that would also meet the criteria, they are rare and the probabilities become vanishingly small. For example, the probability of 3 BL followed by an abnormal is calculated as $P_{3BL\ then\ AB} = P_{BL}^3 \times P_{AB} = 0.1260^3 \times 0.5970 = 0.0012$.

For individuals who are *not truly sensitized* it is also possible to get an abnormal result for the first blood sample, although an abnormal result is less likely. Using the probabilities for single test results that apply to persons not truly sensitized, the probability of falsely identifying an individual is calculated in Table 1 to be 0.0111, or 1.11% (false positive rate). The specificity for this testing plan is 1 – FP = 0.9889, or 98.9% (Middleton et al. 2008).

From Fleiss et al. (2003):

$$PPV = (\text{Sensitivity} \times \text{Prevalence}) / (\text{Sensitivity} \times \text{Prevalence} + (1 - \text{Specificity})(1 - \text{Prevalence}))$$

In order to calculate the PPV for Plan 1, substitute the corresponding values for sensitivity and specificity. A known or assumed value for the prevalence of beryllium sensitization is also necessary for estimating PPV. For example, if we assume a sensitization rate of 4% among a group of exposed workers, then

$$PPV_{\text{Plan 1}} = (0.6817 \times 0.04) / (0.6817 \times 0.04 + (1 - 0.9889)(1 - 0.04)) = 0.719, \text{ or } 71.9\%.$$

If we assume a population prevalence of 2%, the PPV for Plan 1 = 0.556, or 55.6% (Table 4).

Table 1. Likelihood of Meeting Sensitization Criteria of One Abnormal BeLPT (Plan 1), by True Sensitization Status (Middleton et al. 2008)

BeLPT Results That Meet the Criteria: Blood Samples 1	2	3	Probability Calculations[a]	Likelihood of Meeting the Criteria: Truly Sensitized	Not Truly Sensitized
			$P_1 \times P_2 \times P_3$		
AB	-	-	P_{AB}	0.5970	0.0109
BL	AB	-	$P_{BL} \times P_{AB}$	0.0752	0.0002
BL	BL	AB	$P_{BL} \times P_{BL} \times P_{AB}$	0.0095	0.0000
Overall likelihood of meeting the criteria…				0.6817	0.0111

[a] Based on the true status of sensitization, the single test probabilities are:
truly sensitized -------- $P_{AB} = 0.5970$, $P_{BL} = 0.1260$, $P_{NL} = 0.2770$.
not truly sensitized --- $P_{AB} = 0.0109$, $P_{BL} = 0.0158$, $P_{NL} = 0.9733$.

Plan 2. One Abnormal BeLPT and One Borderline BeLPT

Figure 2 illustrates the flow of testing when one abnormal BeLPT and one borderline BeLPT are the minimal criteria for sensitization. If the initial test is abnormal or borderline, another blood sample is taken and split between two laboratories (total of three tests). The *minimum criteria* for sensitization are one abnormal and one borderline, regardless of order. Note that borderline BeLPT results are considered part of the results and the minimum criteria (1 AB and 1 BL) actually exist in the context of three tests: 1 AB, 1 BL, and 1 NL. Of course, more definitive results (e.g., two or three abnormals) are also considered evidence of beryllium sensitization in this plan.

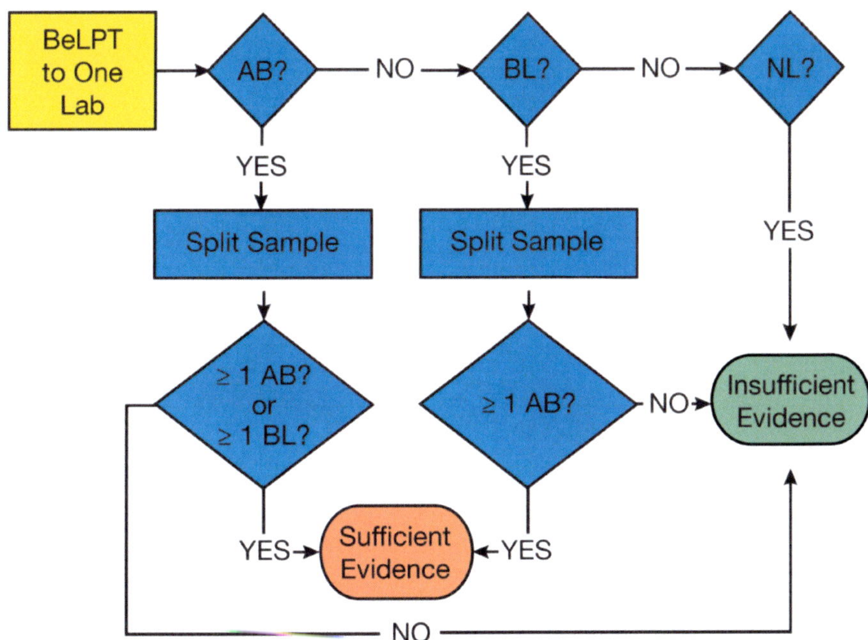

Figure 2. One abnormal BeLPT and one borderline BeLPT provide sufficient evidence for beryllium sensitization (Plan 2) (Middleton and Kowalski 2010).

The sensitivity (0.657) and specificity (0.9992) of Plan 2 are developed in Table 2. Applying the equation for PPV from Fleiss et al. (2003) and assuming a population prevalence of 4%, then

$$PPV_{Plan\ 2} = (0.6599 \times 0.04) / (0.6599 \times 0.04 + (1 - 0.9992)(1 - 0.04))$$
$$= 0.972, \text{ or } 97.2\%.$$

If we assume a population prevalence of 2%, the PPV for Plan 2 = 0.944, or 94.4% (Table 4).

In a previous article (Middleton et al. 2011), we showed that three borderline results could also be considered evidence of sensitization in this testing plan. However, this outcome (3 BL) is rare and contributes little to the plan overall.

Table 2. Likelihood of Meeting Minimal Sensitization Criteria of One Abnormal and One Borderline BeLPT (Plan 2), by True Sensitization Status (Middleton et al. 2008)

BeLPT Results That Meet the Criteria		Probability Calculations[a]	Likelihood of Meeting the Criteria	
Blood Samples			Truly	Not Truly
1	2	$P_1 \times P_2$	Sensitized	Sensitized
AB	AB, AB	$P_{AB} \times (P_{AB} \times P_{AB})$	0.2128	0.0000
	AB, NL	$P_{AB} \times (P_{AB} \times P_{NL} \times 2)$	0.1975	0.0002
	AB, BL	$P_{AB} \times (P_{AB} \times P_{BL} \times 2)$	0.0898	0.0000
	BL, NL	$P_{AB} \times (P_{BL} \times P_{NL} \times 2)$	0.0417	0.0003
	BL, BL	$P_{AB} \times (P_{BL} \times P_{BL})$	0.0095	0.0000
BL	AB, AB	$P_{BL} \times (P_{AB} \times P_{AB})$	0.0449	0.0000
	AB, NL	$P_{BL} \times (P_{AB} \times P_{NL} \times 2)$	0.0417	0.0003
	AB, BL	$P_{BL} \times (P_{AB} \times P_{BL} \times 2)$	0.0190	0.0000
Overall likelihood of meeting the criteria...			0.6569	0.0008

[a] Based on the true status of sensitization, the individual test probabilities are:
truly sensitized -------- $P_{AB} = 0.5970$, $P_{BL} = 0.1260$, $P_{NL} = 0.2770$.
not truly sensitized --- $P_{AB} = 0.0109$, $P_{BL} = 0.0158$, $P_{NL} = 0.9733$.

Plan 3. Two Abnormal BeLPTs

Figure 3 illustrates the flow of testing when two abnormal BeLPT results are required to identify sensitization. If the initial test is abnormal, another blood sample is taken and split between two laboratories (at least three tests total). Borderline results are simply "place-holders" and are repeated. Note that the minimum criteria specified (2 AB) actually exist in the context of three tests (2 AB and 1 NL). Of course, more definitive results (e.g. three abnormal BeLPT results) are also considered evidence of beryllium sensitization in this testing plan.

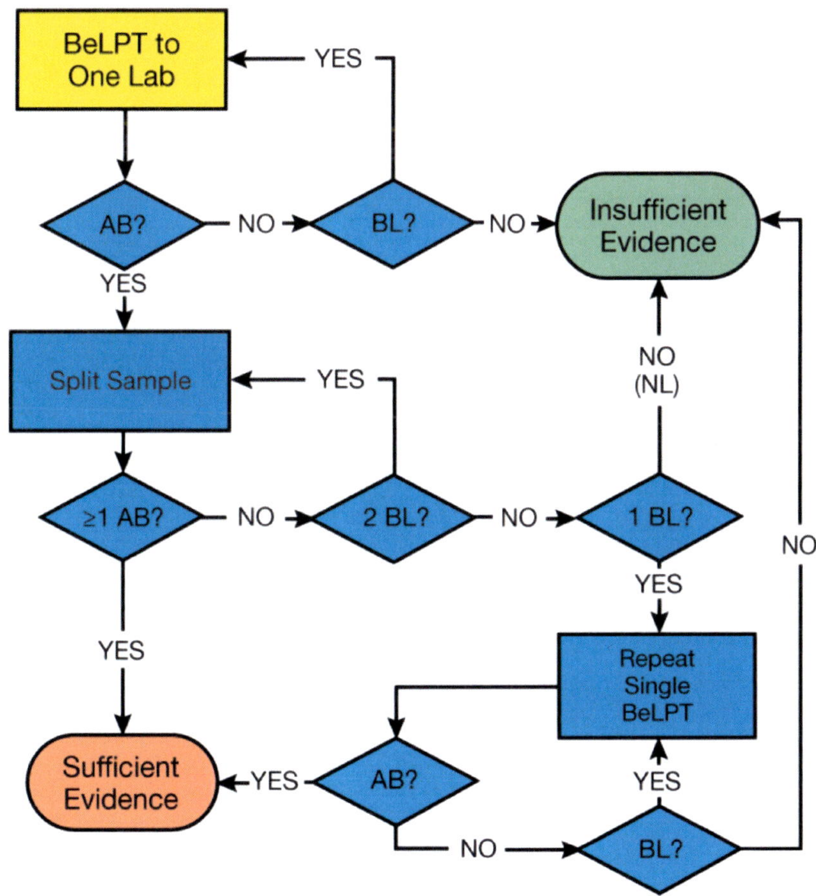

Figure 3. Two abnormal BeLPTs provide sufficient evidence for beryllium sensitization (Plan 3) (Middleton and Kowalski 2010).

As in Plans 1 and 2, we calculate the probabilities for the various outcomes for the two categories (truly sensitized / not truly sensitized) that meet or exceed the minimum criteria for Plan 3 (Table 3). We then sum these probabilities to obtain the sensitivity (0.6115) and specificity (0.9998) for this testing plan.

Table 3. Likelihood of Meeting Sensitization Criteria of Two Abnormal BeLPTs (Plan 3), by True Sensitization Status (Middleton et al. 2008)

BeLPT Results That Meet the Criteria				Probability Calculations[a,b]	Likelihood of Meeting the Criteria	
Blood Samples					Truly Sensitized	Not Truly Sensitized
1	2	3	4	$P_1 \times P_2 \times P_3 \times P_4$		
AB	AB, AB	-	-	$P_{AB} \times (P_{AB})^2$	0.2128	0.0000
	AB, BL	-	-	$P_{AB} \times (P_{AB} \times P_{BL} \times 2)$	0.0898	0.0000
	AB, NL	-	-	$P_{AB} \times (P_{AB} \times P_{NL} \times 2)$	0.1975	0.0002
	BL, NL	AB	-	$P_{AB} \times (P_{BL} \times P_{NL} \times 2) \times P_{AB}$	0.0249	0.0000
		BL	AB	$P_{AB} \times (P_{BL} \times P_{NL} \times 2) \times P_{BL} \times P_{AB}$	0.0031	0.0000
	BL, BL	AB, AB	-	$P_{AB} \times (P_{BL})^2 \times (P_{AB})^2$	0.0034	0.0000
		AB, BL	-	$P_{AB} \times (P_{BL})^2 \times (P_{AB} \times P_{BL} \times 2)$	0.0014	0.0000
		AB, NL	-	$P_{AB} \times (P_{BL})^2 \times (P_{AB} \times P_{NL} \times 2)$	0.0031	0.0000
		BL, NL	AB	$P_{AB} \times (P_{BL})^2 \times (P_{BL} \times P_{NL} \times 2) \times P_{AB}$	0.0004	0.0000
		BL, BL	AB, AB	$P_{AB} \times (P_{BL})^2 \times (P_{BL})^2 \times (P_{AB})^2$	0.0001	0.0000
			AB, BL	$P_{AB} \times (P_{BL})^2 \times (P_{BL})^2 \times (P_{AB} \times P_{BL} \times 2)$	0.0000	0.0000
			AB, NL	$P_{AB} \times (P_{BL})^2 \times (P_{BL})^2 \times (P_{AB} \times P_{NL} \times 2)$	0.0000	0.0000
BL	AB	AB, AB	-	$P_{BL} \times P_{AB} \times (P_{AB})^2$	0.0268	0.0000
		AB, BL	-	$P_{BL} \times P_{AB} \times (P_{AB} \times P_{BL} \times 2)$	0.0113	0.0000
		AB, NL	-	$P_{BL} \times P_{AB} \times (P_{AB} \times P_{NL} \times 2)$	0.0249	0.0000
		BL, NL	AB	$P_{BL} \times P_{AB} \times (P_{BL} \times P_{NL} \times 2) \times P_{AB}$	0.0031	0.0000
		BL, BL	AB, AB	$P_{BL} \times P_{AB} \times (P_{BL})^2 \times (P_{AB})^2$	0.0004	0.0000
			AB, BL	$P_{BL} \times P_{AB} \times (P_{BL})^2 \times (P_{AB} \times P_{BL} \times 2)$	0.0002	0.0000
			AB, NL	$P_{BL} \times P_{AB} \times (P_{BL})^2 \times (P_{AB} \times P_{NL} \times 2)$	0.0004	0.0000
	BL	AB	AB, AB	$P_{BL} \times P_{BL} \times P_{AB} \times (P_{AB})^2$	0.0034	0.0000
			AB, BL	$P_{BL} \times P_{BL} \times P_{AB} \times (P_{AB} \times P_{BL} \times 2)$	0.0014	0.0000
			AB, NL	$P_{BL} \times P_{BL} \times P_{AB} \times (P_{AB} \times P_{NL} \times 2)$	0.0031	0.0000
Overall likelihood of meeting the criteria…					0.6115	0.0002

[a] Based on the true status of sensitization, the individual test probabilities are:
truly sensitized -------- $P_{AB} = 0.5970$, $P_{BL} = 0.1260$, $P_{NL} = 0.2770$.
not truly sensitized --- $P_{AB} = 0.0109$, $P_{BL} = 0.0158$, $P_{NL} = 0.9733$.
[b] The factor "2" was added as needed to consider order (e.g., a, b or b, a).

In order to calculate the PPV for Plan 3, substitute the corresponding values for sensitivity, specificity, and a known or assumed value for the prevalence of beryllium sensitization (Middleton et al. 2008). The sensitivity is 0.6115 and the specificity is 0.9998.

Applying the equation for PPV from Fleiss et al. (2003) and assuming a population prevalence of 4%, we calculate:

$$PPV_{Plan\ 3} = (0.6115 \times 0.04) / (0.6115 \times 0.04 + (1 - 0.9998)(1 - 0.04)) = 0.9922, \text{ or } 99.2\%.$$

If we assume a population prevalence of 2%, the PPV for Plan 3 = 0.984, or 98.4% (Table 4).

COMPARISONS

This discussion compares the relative performance of three plans for testing exposed groups that are asymptomatic. We have calculated estimates for the sensitivity and specificity of each testing plan and showed example calculations for PPV at an assumed prevalence. The sensitivity, specificity, and PPV for a range of population prevalences are shown in Table 4 and in Figure 4. Negative predictive values also vary with population prevalence, but remain high (> 95%) for all three criteria over the range of prevalences shown.

Table 4. Epidemiologic Parameters, by Testing Plan (Middleton et al. 2008)

Plan	Minimum Criteria[a]	Sensitivity / Specificity	Positive Predictive Values (PPV) Assumed Population Prevalences of Be Sensitization						
			1%	2%	3%	4%	5%	7%	10%
1	1 AB	0.682 / 0.9889	0.383	0.556	0.655	0.719	0.764	0.822	0.872
2	1 AB + 1 BL	0.657 / 0.9992	0.893	0.944	0.962	0.972	0.977	0.984	0.989
3	2 AB	0.612 / 0.9998	0.968	0.984	0.990	0.992	0.994	0.996	0.997

[a] Minimum criteria for Plan 1 (1 AB), Plan 2 (1 AB + 1 BL), and Plan 3 (2 AB); results more definitive than the minimum criteria are also acceptable.

The sensitivity and the specificity are constants for each testing plan. Consider a population with a 2% prevalence of "true sensitivity" to beryllium. Plan 1 has a sensitivity of 68.2%, a specificity of 0.9889, and (at 2% prevalence) a positive predictive value of 0.556. While the sensitivity is higher for Plan 1 than for Plan 2 or Plan 3, it has the lowest positive predictive value overall (55.6%) for the three plans. At 2% prevalence, only slightly more than half of the individuals identified as sensitized by Plan 1 are "truly sensitized."

Most practitioners testing asymptomatic persons have insisted on some type of confirmation: some require a second abnormal BeLPT, though others accept a follow-up borderline result as confirmation. Comparing Plans 2 and 3 reveals that Plan 2 is almost as specific as Plan 3, and it is more sensitive (65.7% versus 61.2%). At 2% prevalence, the PPV for Plan 2 (94.4%) approaches that for Plan 3 (98.4%).

The PPV varies across the range of population prevalences, as shown in Table 4. Note that the differences in PPVs for the three plans are larger when the population prevalence is low and smaller when the prevalence is high. Plan 1 is the most sensitive testing plan, but the PPV is only adequate at higher prevalences. The PPV of Plan 2 approaches that of Plan 3 over a range of prevalences, while maintaining a higher sensitivity (0.657 versus 0.612). In most settings the practice of confirming an abnormal result appears beneficial overall, whether with a borderline or an abnormal.

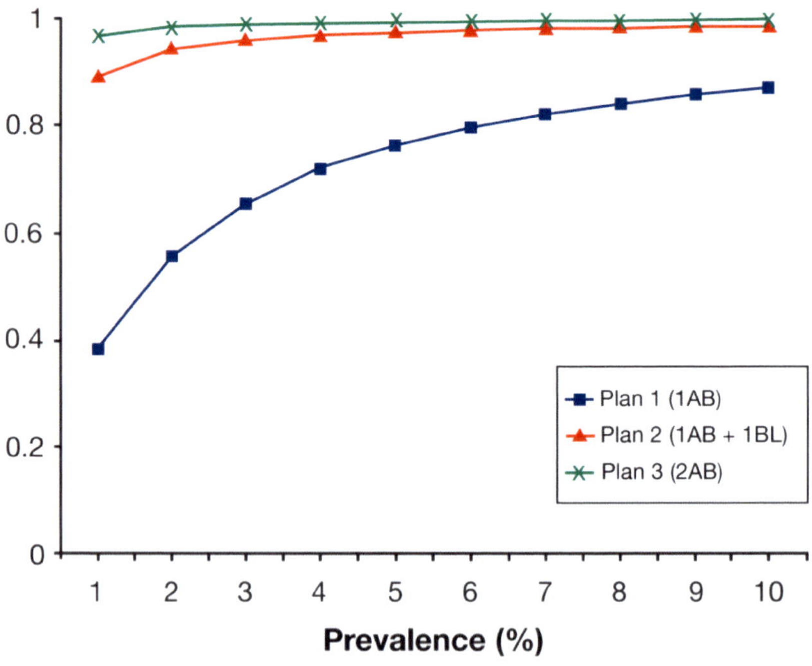

Figure 4. Positive Predictive Value (PPV) by Population Prevalence (Middleton et al. 2008).

INTERPRETATION

Within each testing plan, some results are more definitive predictors than others, but all are considered part of the plan's overall performance. For example if we are employing Plan 2, the minimum criteria (e.g., "1 AB and 1 BL") are nominal outcomes that exist in the testing plan and are a shorthand way of saying: we will accept one abnormal, one borderline, and one normal as sufficient evidence of beryllium sensitization, as well as any results from this testing plan that are more predictive of sensitization.

All of the results shown in Table 2 contribute to the epidemiologic parameters developed for Plan 2 as a whole. The same is true for Plans 1 and 3.

Minimum Criteria and More

The table for each testing plan shows the many ways the criteria for that plan can be satisfied. Our intent has been to characterize the testing plans as they are employed and compare their performance overall. These calculations are more useful for designing programs for group screening than for assigning probabilities retrospectively to specific outcomes.

In a recent paper, we did explore some specific outcome combinations in the context of Plan 2 (Middleton et al. 2011). The minimum three-result criteria (1 AB + 1 BL + 1 NL) define a plan that includes this combination and more definitive combinations; overall, this plan functions well across a range of prevalences. The individual three-result combination (1 AB + 1 BL + 1 NL) has lower positive predictive values than other combinations in the testing plan (e.g., 3 AB); this is to be expected, given that the minimum criteria (1 AB + 1 BL + 1 NL) is the least definitive outcome combination for confirming immunologic sensitization in Plan 2.

STRENGTHS AND WEAKNESSES

The estimates we have made are based on work by Stange et al. (2004) using data from a large group of exposed workers. Without an absolute "gold standard" we cannot fully evaluate the *sensitivity* of the BeLPT and must refer to the logic present in Stange et al. (2004) for support. However, it is possible to look at the *specificity* of the test in groups not exposed by occupational history and therefore presumably not truly sensitized to beryllium. The examples below suggest that the BeLPT has a high specificity, consistent with Stange's estimate for this parameter (96.9%).

One Abnormal BeLPT Result

Yoshida et al. (1997) tested 159 new hires in Japan. The researchers stated that there was no prior exposure or evidence of beryllium disease in the group. Of these apparently unexposed new hires, two had an abnormal result. No information about follow-up testing for confirmation was published.

Stange et al. (2004) tested new hires that were not included in the database used for probability estimates. The researchers used interviews and questionnaires to identify 291 new hires who had no known exposure to beryllium. Three of the 291 unexposed workers had one abnormal BeLPT on the first round of testing.

Cummings et al. (2007) reported four abnormal results among 97 pre-hires hired between 2000 and 2004. Follow-up testing occurred after the employee began work. That is, they were no longer unexposed by occupational history at follow-up.

Thomas et al. (2009) reported one abnormal result among 82 new hires. The authors state that none of these workers were known to have been exposed to beryllium before they were hired. Follow-up testing occurred after the employee began work.

Two Abnormal BeLPT Results

Stange et al. (2004) also reported the follow-up testing of the three individuals with abnormal results among the 291 new hire results; they found none of them to be sensitized to beryllium when sensitization was defined as 2 AB.

Silveira et al. (2003) reported BeLPT results on 527 pre-hires from three separate communities. Detailed exposure histories were taken and the blood was obtained prior to any occupational exposure to beryllium. One individual was sensitized (2 AB), but may have had prior occupational exposure. Considering this information and other unpublished data, the authors concluded that if beryllium sensitization without occupational exposure occurs, it must be rare.

Limitations

We have estimated the epidemiological characteristics of three testing plans used to test exposed workers for immunologic sensitivity to beryllium. While these plans are clearly specified in this chapter, real world screening programs are not always so well-defined or so closely followed. Because of this, we must acknowledge that these three plans are somewhat idealized models of BeLPT testing. Further, the characteristics of the three plans do not represent the minimal characteristics that define them, but apply overall to the multiple outcomes in each plan accepted as sufficient evidence of sensitization.

Further, there are a number of assumptions implicit to these three testing plans described by Middleton et al. (2006, 2008). BeLPT results from various laboratories are considered interchangeable in these calculations. Beryllium sensitization is considered to be a relatively constant state over the testing period. T-lymphocytes from all sensitized persons are assumed to respond similarly to in vitro challenges by beryllium sulfate, as are the lymphocytes of all persons not sensitized. These assumptions are not always completely met.

The estimates of BeLPT performance by Stange et al. (2004) are based on data collected between 1992 and 2001. We cannot be sure that laboratory performance has remained the same. We note that our work builds on that of Stange et al. (2004), ultimately depending on the data he used. Finally, we note that it isn't possible to fully evaluate the epidemiologic characteristics that Stange et al. (2004) provided for a single BeLPT in the absence of a gold standard for beryllium sensitization.

CONCLUSION

Subject matter experts and even government agencies differ in their use of the BeLPT. The U.S. Department of Energy has a comprehensive program for preventing exposure to beryllium and provides testing for both current and former workers who may have been exposed (U.S. Department of Energy, 1999). The Department of Defense has discouraged use of the BeLPT for screening or surveillance of exposed workers (U.S. Department of the Air

Force, 2002; U.S. Department of the Army, 2002a, b). Recently, the U.S. Air Force asked the National Research Council (NRC) to conduct an independent review of the scientific literature on beryllium. The 2008 NRC report discussed the role of the BeLPT in worker surveillance, concluding that:

> "The BeLPT is integral to any screening program. No alternative tests have been adequately validated to be put into practice outside research settings."

The testing algorithm recommended to the Air Force was taken from a manuscript by Middleton et al. (2008); this algorithm corresponds to Plan 2 in this chapter (1 AB + 1 BL). The NRC report also noted that current practice is not always limited to a well-defined algorithm, but considers all BeLPT results available for the individual tested. The committee acknowledged some uncertainty, but suggested that this practice continue until data are available to clarify this issue (NRC 2008).

Citing Stange et al. (2004) and Middleton et al. (2006; 2008), the committee also advised consideration of the *prevalence* of beryllium sensitization in the population tested. We agree and we add that the information in this chapter is more useful when the testing plans can be compared at specific prevalences.

The information in this chapter was developed to facilitate more objective discussions of testing and the design of testing plans. We note as a precaution that the parameters developed in this chapter apply to the three respective testing plans as a whole and do not represent the sensitivity, specificity, or positive predictive values of the individual minimal criteria alone.

Finally, it is also important to note that the analyses presented here only attempt to model testing designed for persons who do not have symptoms associated with CBD. When other factors are present, such as respiratory signs or symptoms that increase the likelihood of CBD, clinical considerations may dictate an entirely different approach to diagnosis and treatment.

ACKNOWLEDGMENTS

We wish to acknowledge our debt to the work of Dr. Arthur W. Stange and co-authors Dr. F. Joseph Furman and Dr. Duane E. Hilmas in their manuscript "The beryllium lymphocyte proliferation test: Relevant issues in beryllium health surveillance" (Stange et al. 2004). We also acknowledge the advice and contributions of many distinguished individuals in government and academia, many of whom participated as co-authors in our previous publications on this topic (Middleton et al. 2006; 2008; 2010; 2011). Finally, we thank Carolyn Srite for proofreading and formatting support in preparing this chapter for publication.

REFERENCES

Agency for Toxic Substances and Disease Registry (ATSDR). Toxicological Profile for Beryllium. 2002. Available from: http://www.atsdr.cdc.gov/toxprofiles/tp4.pdf.

Bailey, RL; Thomas, CA; Deubner, DC; Kent, MS; Kreiss, K; Schuler, CR. (2010). Evaluation of a preventive program to reduce sensitization at a beryllium metal, oxide, and alloy production plant. *J. Occup. Environ. Med*, 52(5), 505-512.

Cummings, KJ; Deubner, DC; Day, GA; Henneberger, PK; Kitt, MM; Kent, MS; Kreiss, K; and Schuler, CR. (2007). Enhanced preventive programme at a beryllium oxide ceramics facility reduces beryllium sensitization among new workers. *Occup. Environ. Med*, 64(2), 134-140.

Deubner, DC; Goodman, M; Iannuzzi, J. (2001). Variability, predictive value, and uses of the beryllium blood lymphocyte proliferation test (BLPT): Preliminary analysis of the ongoing workforce survey. *Appl. Occup. Environ. Hyg*, 16, 521–526.

Eisenbud, M; Lisson, J. (1983 March). Epidemiological aspects of beryllium-induced nonmalignant lung disease: a 30-year update. *J. Occup. Med*, 25(3), 196–202. (PubMed)

Fleiss, JL; Levin, B; Paik, MC. (2003). Statistical methods for rates and proportions. 3rd edition. New York, NY: John Wiley & Sons.

Hardy, HL; Tabershaw, IR. (1946). Delayed chemical pneumonitis occurring in workers exposed to beryllium compounds. *J. Indust. Hyg. & Toxicol*, 28: 197-211.

Henneberger, PK; Goe, SK; Miller, WE; Doney, B; Groce, DW. (2004). Industries in the United States with airborne beryllium exposure and estimates of the number of current workers potentially exposed. *J. Occup. Environ. Hyg*, 1(10), 648-659.

Kreiss, K; Newman, LS; Mroz, MM; Campbell, PA. (1989 July). Screening blood test identifies subclinical beryllium disease. *J. Occup. Med*, 31(7), 603–608. (PubMed)

Maier, LA. (2001). Beryllium health effects in the era of the beryllium lymphocyte proliferation test. *Appl. Occup. Environ. Hyg*, 16, 514-520.

Maier, LA; Martyny, JL: Liang, J; Rossman MD. (2008). Recent chronic beryllium disease in residents surrounding a beryllium facility. *Am. J. Respir. Crit. Care Med*, 177(9), 1012-1017.

Middleton, DC. (1998). Chronic beryllium disease: Uncommon disease, less common diagnosis. *Environ. Health Perspect*, 106, 765–767.

Middleton, DC; Lewin, MD; Kowalski, PJ; Cox, SS; Kleinbaum D. (2006). The BeLPT: algorithms and implications. *Am. J. Ind. Med*, 49, 36–44.

Middleton, DC; Fink, J; Kowalski, PJ; Lewin, MD; Sinks, T. (2008). Optimizing BeLPT criteria for beryllium sensitization. *Am. J. Ind. Med,* 51, 166–172.

Middleton, D; Kowalski, P. (2010). Advances in Identifying Beryllium Sensitization and Disease. *Int. J. Environ. Res. Public Health*, 7, 115-124.

Middleton, DC; Mayer, AS; Lewin, MD; Mroz, MM; Maier LA. (2011). Interpreting borderline BeLPT results, *Am. J. Ind. Med*, 54, 205-209.

Mroz, MM; Kreiss, K; Lezotte, DC; Campbell, PA; Newman, LS. (1991). Reexamination of the blood lymphocyte transformation test in the diagnosis of chronic beryllium disease. *J. Allergy Clin. Immunol*, 88, 54-60.

Newman, LS; Mroz, MM; Balkissoon, R; Maier, LA. (2005 January 1). Beryllium sensitization progresses to chronic beryllium disease: a longitudinal study of disease risk. *Am. J. Respir. Crit. Care Med.*, 171(1), 54-60.

NRC (National Research Council) of the National Academies. (2008). Managing health effects of beryllium exposure. Washington, DC: The National Academies Press; Washington, DC, p. 168.

Richeldi, L; Kreiss, K; Mroz, MM; Zhen, B; Tartoni, P; Saltini, C . (1997). Interaction of genetic and exposure factors in the prevalence of berylliosis. *Am. J. Ind. Med*, 32(4), 337-340.

Rossman, MD; Stubbs, CW; Lee, E; Aryris, E; Magira, E; and Monos, D. (2002). Human leukocyte antigen class II amino acid epitopes: Susceptibility and progression markets for beryllium hypersensitivity. *Am. J. Respir. Crit. Care Med*, 165(6), 788-794.

Silveira, L; Bausch, M; Mroz, M; Maier, L; and Newman, L. (2003). Beryllium sensitization in the "general population." *Sarcoidosis Vasc. Diffuse Lung Dis*, 20, 157.

Silveira, LJ; McCanlies, EC; Fingerlin; TE; Van Dyke, MV; Mroz, MM; Strand, M; Fontenot, AP; Bowerman, N.; Dabelea, DM; Schuler, CR;

Weston, A.; and Maier, LA. (2012). Chronic Beryllium Disease, HLA-DPB1, and the DP Peptide Binding Groove. *J. Immun*, 189(8), 4014-4023.

Stange, AW; Furman, FJ; Hilmas, DE. (2004). The beryllium lymphocyte proliferation test: relevant issues in beryllium health surveillance. *Am. J. Ind. Med*, 46, 453–62.

Thomas, CA; Bailey, RL; Kent, MS; Deubner, DC; Kreiss, K; Schuler, CR. (2009). Efficacy of a program to prevent beryllium sensitization among new employees at a copper-beryllium alloy processing facility. *Public Health Rep*, 124(Suppl 1), 112-124

U.S. Department of the Air Force. 2002. Beryllium surveillance and medical monitoring policy (memorandum of 28 August 2002).

U.S. Department of the Army. 2002a.Work place exposure to beryllium (memorandum of 7 May, AMCSG-I, 40-5e).

U.S. Department of the Army. 2002b. Beryllium surveillance and medical monitoring policy (memorandum of 15 August, DASG-PPM-NC).

U.S. Department of Energy. 1999. Chronic Beryllium Disease Prevention Program; Final Rule. 10 CFR 850.

Yoshida, T; Shima, S; Nagaoka, K; Taniwaki, H; Wada, A; Kurita, H, Morita, K. (1997 July). A study on the beryllium lymphocyte transformation test and the beryllium levels in working environment. *Ind. Health*, 35, 374–379.

In: Beryllium
Editor: Pauleen Dyer
ISBN: 978-1-63321-590-0

Chapter 2

CLINICAL ASPECTS OF CHRONIC BERYLLIUM DISEASE: A REVIEW

Hans Schweisfurth, M.D.*

Director of the Institute for Pulmonary Research (IPR)
Cottbus, Germany

ABSTRACT

Chronic beryllium disease (CBD) can occur by exposure to beryllium, a light earth alkaline metal. Beryllium and its compounds are used in the electronic-, computer-, atomic-, and ceramics industries. Beryllium is also found in fossil fuels and can be released through combustion. The uptake of beryllium and its compounds occur mainly by inhalation. Acute beryllium disease (ABD) is a toxic chemical reaction to beryllium and its compounds and mainly effects the respiratory tract. CBD is a specific cell-mediated delayed immune response on beryllium. The susceptibility of CBD is associated with the HLA-DPB1 gene possessing glutamic acid at 69th position (E69) of the ß-chain. Pathophysiologically non-necrotizing granulomas are detectable most commonly in lung and skin. The main symptoms include dry cough, fatigue, weight loss and increasing shortness of breath. CBD can be mistaken for sarcoidosis due to the similar symptoms and the formation of granulomas. However, the beryllium lymphocyte proliferation test in

* Correspondence: Hans Schweisfurth, Professor of Medicine; Institute for Pulmonary Research (IPR), Walther-Rathenau-Strasse 11, D-03044 Cottbus, Germany, Email: ipr@pulmologisches-forschungsinstitut.de.

CBD is positive in contrast to sarcoidosis. Therapeutic steroids are used in addition to avoidance of beryllium exposure. Even if there is no further exposure, CBD can also progress over decades and end in pulmonary fibrosis. Beryllium and its compounds are considered to be carcinogenic, because in patients suffering from ABD or CBD and in individuals exposed to beryllium an increased rate of lung cancer was observed.

INTRODUCTION

The relatively rare chemical element beryllium was isolated from the mineral beryl in 1798 [1]. In the periodic table, it has the atomic number 4 with an atomic weight of 9.012. Beryllium is one of the lightest metals with a low density of 1.846 g/cm^3 at 20 ° C and possesses a high melting point of 1.287 ° C [2, 3]. It occurs naturally only with other elements in minerals. Beryllium is three times lighter than aluminum and has excellent electrical and thermal conductivity and is not magnetic. It is insoluble in water but soluble in acids and alkalis. Due to its low density beryllium is relatively transparent to X-rays. About more than 40 beryllium-bearing minerals are known, but only Beryl ($3BeO \cdot Al_2O_3 \cdot 6SiO_2$) and Bertrandite ($4BeO \cdot 2SiO_2 \cdot H_2O$) are commercially important [2, 3]. Beryl is found in China, Brazil and the former Soviet Union. Bertrandite is mostly mined in the U.S. (Utah, Nevada). Beryllium is used in many industrial products (Table 1). Fossil fuels such as coal and oil contain also beryllium compounds. Beryllium can also be released by incineration plants.

Beryllium is considered to be carcinogenic, because in patients suffering from acute (ABD) or chronic beryllium disease (CBD) or in workers exposed to beryllium an increased rate of lung cancer was found [2, 3, 4, 5, 6]. Workers with high exposure to beryllium (> 10 µg/m^3) have also a higher rate of carcinoma of the urinary tract [7]. However, in animals mutagenic and cytotoxic effects of beryllium were not clearly detected and remain under discussion [8, 9].

PREVALANCE

The highest levels of human exposure to beryllium may occur through inhalation of beryllium dust, dermal contact or by retained beryllium-containing foreign bodies [2, 10]. Beryllium miners, beryllium alloy makers and fabricators, ceramics workers, missile technicians, nuclear reactor workers, electric and electronic equipment workers and jewelers have a high potential for exposure [2, 3, 11]. ABD is nowadays very rare. However, 3 % - 16 % of the beryllium exposed develop CBD [12, 13, 14, 15]. In a ceramics plant, the overall prevalence of CBD was 5.3 % [16].

In the U.S. it is estimated that approximately 134,000 to 200,000 current and at least 800,000 to 1 million total individuals have been exposed to beryllium who are potentially at risk for developing CBD [17, 18, 19, 20].

In 1952 the U.S. Beryllium Case Registry was established at the Massachusetts Institute of Technology. This registry later moved to the National Institute for Occupational Safety and Health and is now closed. Approximately 900 cases were entered into the registry, including 65 cases among family members exposed to dust taken home by the workers and residents exposed through off-site air pollution [21]. Data suggest that early detection identifies subjects who are sensitized to beryllium and that these individuals are at risk for progressing into CBD, which can be developed in approximately 30 % [22, 23].

ACUTE BERYILLIUM DISEASE

ABD is defined as an irritative direct chemical-toxic reaction of the lung related to high exposure of beryllium with less than one year duration [2, 24, 25]. ABD may be fatal in 10 % of the patients [2]. The first acute beryllium intoxication was described in Germany in 1933 and in the U.S. in 1936 [26, 27]. In 1945 more cases were reported [28]. The clinical features of ABD are similar to acute sarcoidosis. Soluble beryllium compounds cause irritation of the respiratory tract such as rhinitis, pharyngitis, tracheitis, bronchitis and alveolitis. Further symptoms are dyspnea, fatigue, fever, night sweats, cough and hemoptysis.

The symptoms begin usually within hours or a few days after beryllium exposure. Pulmonary function tests show obstructive or restrictive ventilation disorder with limited gas exchange. Radiographically diffuse interstitial pulmonary infiltrates are detectable. Lung biopsies show features of a lymphocytic interstitial pneumonitis.

CHRONIC BERYILLIUM DISEASE

CBD was described for the first time in 1946 as a delayed chemical pneumonitis in 17 patients, who worked in the fluorescent lamp industry. The patients suffered from weight loss, shortness of breath during physical exertion and cough about 6 months after the first exposure [29]. In 1962, very high beryllium concentrations in lung tissue were detected in CBD whereas no beryllium was found in patients with sarcoidosis [30].

IMMUNOPATHOGENESIS

CBD, formerly known as "berylliosis" or "chronic berylliosis", is an inflammatory lung disease caused by inhalation exposure of soluble or insoluble forms of beryllium, which induces a cell-mediated delayed immune response with formation of granulomas and varying degrees of interstitial fibrosis [2]. The chemical behavior of beryllium is largely a result of its small atomic and ionic radii. Therefore, it is too small to be an antigenic, it may function as a hapten, binding with a large carrier molecule to form an antigen. Beryllium acts as a major histocompatibility complex class (MHC) II-restricted antigen, stimulating local proliferation and accumulation of beryllium-specific CD4(+) T cells [31]. Endogenous peptides bind MHC II and beryllium, forming a complex recognized by pathogenic CD4(+) T cells [32]. In patients with CBD pathogenic CD4(+) T cells have been found mainly in BAL [33, 34]. Recently it has been shown that a dysfunctional pathway of cytotoxic T-lymphocyte antigen-4 contributes to persistent inflammation in CBD [35]. The severity of CBD was inversely correlated with the frequency of tetramer-binding CD4(+) T cells in the lung [36].

The susceptibility of CBD is associated with the HLA-DPB1 gene possessing glutamic acid at 69th position (E69) of the ß-chain in 84 % to 97 % of the patients [37, 38, 39, 40, 41, 42]. Less-frequent E69 variants might be associated with greater risk of CBD [43]. In addition to HLA-DPB1 other genes and environmental factors could be responsible for the manifestation of CBD [44, 45, 46, 47].

Clinical Features

CBD can develop after beryllium exposure over years in 1 % to 5 % [33, 34]. The average latency of the manifestation of CBD is over 10 years. It is reported that CBD can occur after an exposure of more than 30 years [48]. Mostly the manifestation of CBD begins beyond the 40th year of life, in contrast to sarcoidosis, which mainly appears at younger age (Table 2). It is still unknown, if cessation of exposure in sensitized individuals reduces the progression rate to CBD [49].

The history of CBD can be completely asymptomatic. Restrictions on lung function and gas exchange especially during physical exertion are detectable at the early stage. Dry cough, fatigue, chest pain, weight loss, night sweats, fever and loss of appetite can occur. In rare cases other organs such as liver, spleen, heart, skeletal muscles, pancreas or bones are affected.

Diagnosis

By chest radiography small parenchymal nodules or large conglomerates with preference of the upper lobe can be detected. Enlarged mediastinal lymph nodes are present in about one-third of the patients. However, the conventional chest radiography can be without pathological findings. A higher sensitivity can be reached by thin-section CT. The most common CT abnormalities were parenchymal nodules and septal lines. By the use of CT, abnormalities were detected in 77 % of patients with normal radiographs [50].

Pathohistologically a CD4(+) T-cells alveolitis and non-necrotizing granulomas can be found within the lung, but also granulomas are detectable in skin, liver, spleen and myocardium and may be misdiagnosed as sarcoidosis [33, 51, 52, 53, 54, 55, 56]. Beryllium-induced oxidative stress plays a role in the pathogenesis of granulomatous inflammation in CBD and also mast cells may contribute to the development of fibrosis, because they produce a 17.8-kDa isoform of basic fibroblast growth factor, which is involved in regulating fibroblast proliferation [57].

It has been known since 1951 that patients develop a delayed skin reaction by beryllium compounds. Therefore patch and intracutaneous skin tests are used for screening [58, 59]. As lymphocytes in blood proliferate after the addition of beryllium sulfate, this property is used as a diagnostics tool in the beryllium lymphocyte proliferation test (BeLPT). The proliferation rate is measured by absorption of radiolabeled thymidine in the lymphocytes. Unfortunately, the BeLPT is not standardized with the consequence that an abnormal test result should be confirmed [60]. The BeLPT with lymphocytes from bronchoalveolar lavage (BAL) can occur false negative in cigarette smokers [2, 61]. The combination of a beryllium exposure history with evidence of granulomatous inflammation on lung biopsy and a positive BAL-BeLPT makes the diagnosis of CBD reasonably secure, but also the clinical symptoms (Table 3) should be considered [62].

The sensitivity of the BeLPT varies between 38 % and 100 % [14, 63, 64]. As there is a small cross-laboratory reproducibility there will be cases of CBD, which cannot be detected on the basis of false negative test results [65]. However, the BeLPT for the diagnosis of CBD is generally accepted because the BeLPT in sarcoidosis and other granulomatous diseases is negative [14, 59]. A follow-up study of patients with sarcoidosis showed that 17 out of 121 patients had also beryllium exposure at work or at home, but none of these sarcoidosis patients had a positive BeLPT [66].

In another study, in 84 patients seen for re-evaluation or making a diagnosis of sarcoidosis, beryllium exposure was recognized and a diagnosis of CBD was made in 34, which had a positive BeLPT [67].

Not every beryllium exposed individual with a positive BeLPT suffers from CBD [15]. The beryllium sensitivity is found in 1 % to 16 % of exposed workers [41]. In the nuclear and conventional munition industry 1.4 % - 2.3 % of workers were sensitized to beryllium but only 0.2 % suffered from CBD [68, 69, 70]. The risk of beryllium sensitization depends on the genotype [71, 72]. Some beryllium exposed had repeated positive test results as an indication of a beryllium sensitization but they did not have any clinical symptoms or granulomas. A longitudinal study showed that 31 % of beryllium sensitized individuals developed CBD after an average of 3.8 years. This corresponds to a disease rate of 6 % to 8 % per year. There was not found any difference in terms of average age, sex, ethnicity, smoking status, or beryllium exposure time between those who progressed to CBD and those who remained sensitized without disease [22].

The detection of beryllium in the granulomas can support the diagnosis of CBD, but very low beryllium tissue concentrations are difficult to measure [73, 74]. Beryllium is also detectable in urine. A field study revealed that there was a higher beryllium excretion rate in the urine of the beryllium exposed compared to the non-exposed. A beryllium air concentration of 0.2 $\mu g/m^3$ responds to a urinary beryllium excretion of 0.15 µg/l [75]. Therefore it is recommended, that in beryllium exposed workers beryllium measurements should routinely be performed in urine [76].

THERAPY

Patients at an early stage (beryllium sensitized and granulomas) of CBD without any symptoms or pulmonary function deterioration, should be regularly monitored (pulmonary function tests, physical stress test and chest radiograph). Serological markers of disease activity, as they are also used in sarcoidosis (soluble interleukin 2 receptor, angiotensin-converting enzyme and neopterin), may also indicate the inflammatory activity in CBD [54, 77, 78].

A corticosteroid therapy is indicated for progressive disease. Oral doses of prednisolone of 0.5 - 0.8 mg/kg body weight daily are recommended. Corticosteroids stabilize or improve most patients with CBD. Relapses under dose reduction can occur. Long-lasting remissions and refractory processes are also described. The response to long-term corticosteroids in CBD is variable like that in sarcoidosis. Significant improvement of lung function may follow after cessation of beryllium exposure [79].

There are no systematic studies concerning the use of other immunosuppressive drugs. For patients who do not respond to high doses of prednisolone, a second-line therapy with cyclophosphamide, hydroxylchloroquine, methotrexate, azathioprine or cyclosporine should be tried out [77, 80, 81, 82]. Infliximab, a monoclonal antibody, is successfully used in severe cases of sarcoidosis, but in CBD the application has not sufficiently been investigated [83, 84, 85]. Adjuvant therapies have not been specifically studied in patients with CBD, but supplemental oxygen, bronchodilators, vaccination to prevent influenza and pneumococcal pneumonia and pulmonary rehabilitation should be applied [85]. In end-stage cases, which are relatively rare, lung transplantation should be considered [77].

PREVENTION

Beryllium and its compounds are an important but preventable cause of chronic lung disease and worldwide large numbers of workers are exposed, but there are no international exposure standards. A daily weighted average

(DWA) exposure to 0.02 mg/m³ for 24 hours was classified as harmless [12]. This standard is accepted by the Environmental Protection Agency of the U.S. [86]. The time weighted average (TWA) for 8 hours exposure in the workplace are regionally different. A study showed that exposure to TWA values below 0.02 mg/m³ have no health effects [87]. However the Occupational Safety and Health Administration (OSHA) of the U.S. determined the permissible exposure limit (PEL) of 2.0 μg/m³ air for an 8-hour work day, but also this concentration will not sufficiently protect workers so that a new standard will be evaluated by OSHA [20].

Therefore the strict hygienic standards should be followed [88]. Reduction of beryllium exposure improved gas exchange and reversed radiographic abnormalities in some beryllium workers [89]. Occupational health monitoring using BeLPT and a posteroanterior chest radiograph with pulmonary function test are recommended for all beryllium exposed [90]. Genetic screening programs may become reliable tools to prevent CBD [91]. The combination of respiratory and dermal protection, improved ventilation, dust migration control and education of workers and management are very helpful measures to reduce beryllium sensitization successfully [92].

Table 1. Application of beryllium and its compounds in the industry [2, 3, 77]

- Construction material in alloys for aircrafts
- Reflector for neutrons in reactors
- Neutron multiplier in 'Fast breeder reactors'
- Windows in X-ray tubes
- MRI equipment
- Spring tools
- Golf clubs
- Brake discs for space shuttle
- Relay contacts
- Electronics and computer industry
- Ceramics and jeweler industry
- Dental technology

Table 2. Differential characteristics between CBD and sarcoidosis [93].

	CBD	**Sarcoidosis**
Course	Progressive	Acute or chronic
Sex preference	Non	Female ?
Age	> 40 years	< 40 years
Non-necrotizing granulomas	++	++
Lung	++	++
Skin	+	+
Erythema nodosum	-	+
Eyes	+	++
Liver	+	++
Heart	+	++
Nervous system	?	++
BAL	CD4/CD8 elevated	CD4/CD8 elevated
BeLPT in blood and BAL	Positive	Negative

Table 3. Criteria for diagnosing CBD [2, 48, 86].

1) History of beryllium exposure.
2) Histopathological evidence of non-caseating granulomas or mononuclear cell infiltrates in the absence of infection.
3) Positive blood- or BAL-BeLPT.
4) Clinical findings such as respiratory symptoms, reticular pulmonary infiltrates, obstructive or restrictive ventilation disorder, decreased diffusion capacity or impaired gas exchange during physical exercise.

REFERENCES

[1] Vauquelin L-N. De l'Aiguemarine, ou Béril; et découverie d'une terre nouvelle dans cette pierre. *Annales de Chimie* 1798; 26: 155–169.

[2] WHO. Concise international chemical assessment document 32. Beryllium and beryllium compounds, Geneva 2001.

[3] Report on carcinogens, U.S. Department of Health and Human Services, Public Health Service, National Toxicology Program, 12th edition 2011; 67-70.

[4] Sanderson WT, Ward EM, Steenland K, Petersen MR. Lung cancer case-control study of beryllium workers. *Am. J. Ind. Med.* 2001; 39: 133-144.

[5] Steenland K, Ward E. Lung cancer incidence among patients with beryllium disease: a cohort mortality study. *J. Natl. Cancer Inst.* 1991; 83: 1380-1385.

[6] Ward E, Okun A, Ruder A, Fingerhut M, Steenland K. A mortality study of workers at seven beryllium processing plants. *Am. J. Ind. Med.* 1992; 22: 885-904.

[7] Schubauer-Berigan MK, Couch JR, Petersen MR, Carreón T, Jin Y, Deddens JA. Cohort mortality study of workers at seven beryllium processing plants: update and associations with cumulative and maximum exposure. *Occup. Environ. Med.* 2011; 68: 345-353.

[8] Strupp C. Beryllium metal I. Experimental results on acute oral toxicity, local skin and eye effects, and genotoxicity. *Ann. Occup. Hyg.* 2011; 55: 30-42.

[9] Strupp C. Beryllium metal II. A review of the available toxicity data. *Ann. Occup. Hyg.* 2011; 55: 43-56.

[10] Fireman E, Shai AB, Lerman Y, Topilsky M, Blanc PD, Maier L, Li L, Chandra S, Abraham JM, Fomin I, Aviram G, Abraham JL. Chest wall shrapnel-induced beryllium *Sarcoidosis Vasc. Diffuse Lung Dis.* 2012; 29:147-150.

[11] Kreiss K, Day GA, Schuler CR. Beryllium: a modern industrial hazard. *Annu. Rev. Public Health* 2007; 28: 259-277.

[12] Kreiss K, Mroz MM, Newman LS, Martyny J, Zhen B. Machining risk of beryllium disease and sensitization with median exposures below 2 micrograms/m^3. *Am. J. Ind. Med.* 1996; 30: 16-25.

[13] Kreiss K, Mroz MM, Zhen B, Martyny JW, Newman LS. Epidemiology of beryllium sensitization and disease in nuclear workers. *Am. Rev. Respir. Dis.* 1993; 148: 985–991.

[14] Kreiss K, Wasserman S, Mroz MM, Newman LS. Beryllium disease screening in the ceramics industry. Blood lymphocyte test performance and exposure-disease relations. *J. Occup. Med.* 1993; 35: 267–274.

[15] Kreiss K, Newman LS, Mroz MM, Campbell PA. Screening blood test identifies subclinical beryllium disease. *J. Occup. Med.* 1989; 31: 603–608.

[16] Henneberger PK, Cumro D, Deubner DD, Kent MS, McCawley M, Kreiss K. Beryllium sensitization and disease among long-term and short-term workers in a beryllium ceramics plant. *Int. Arch. Occup. Environ. Healt*h 2001; 74: 167–176.

[17] Cullen MR, Cherniack MG, Kominsky JR. Chronic beryllium disease in the United States. *Sem. Respir. Med.* 1986; 7: 203-209.

[18] Falta MT, Bowerman NA, Dai S, Kappler JW, Fontenot AP. Linking genetic susceptibility and T cell activation in beryllium-induced disease. *Proc. Am. Thorac. Soc.* 2010; 7: 126-129.

[19] Henneberger PK, Goe SK, Miller WE, Doney B, Groce DW. Industries in the United States with airborne beryllium exposure and estimates of the number of current workers potentially exposed. *J. Occup. Environ. Hyg.* 2004; 1: 648–659.

[20] Sawyer RT, Maier LA. Chronic beryllium disease: an updated model interaction between innate and acquired immunity. *Biometals* 2011; 24: 1-17.

[21] Middleton D, Kowalski P. Advances in identifying beryllium sensitization and disease. *Int. J. Environ. Res. Public Health* 2010; 7: 115-124.

[22] Newman LS, Mroz MM, Balkissoon R, Maier LA. Beryllium sensitization progresses to chronic beryllium disease: a longitudinal study of disease risk. *Am. J. Respir. Crit. Care Med.* 2005; 171: 54-60.

[23] Newman LS, Lloyd J, Daniloff E. The natural history of beryllium sensitization and chronic beryllium disease. *Environ. Health Perspect.* 1996; 104(Suppl 5): 937-943.

[24] Cummings KJ, Stefaniak AB, Virji MA, Kreiss K. A reconsideration of acute beryllium disease. *Environ. Health Perspect.* 2009; 117: 1250-1256.

[25] Hooper WF. Acute beryllium lung disease. *N C Med. J.* 1981; 42: 551-553.

[26] Gelman I. Poisoning by vapors of beryllium oxyfluoride. *J. Indust. Hyg. Toxicol.* 1936; 18: 371-379.

[27] Weber HH, Engelhardt WE. Über einen Apparat zur Erzeugung niedriger Staubkonzentrationen von grosser Konstanz und eine Methode zur mikrogravimetrischen Staubbestimmung. Anwendung bei der Untersuchung von Stäuben aus der Beryllium Gewinnung. *Zbl. Gewerbehyg. Unfallverh.* 1933; 10: 41-47.

[28] Van Ordstrand HS, Hughes CR, DeNardi JM, Carmody MG. Beryllium poisoning. *JAMA* 1945; 129:1084–1090.

[29] Hardy HL, Tabershaw IR. Delayed chemical pneumonitis occurring in workers exposed to beryllium compounds. *J. Ind. Hyg. Toxicol.* 1946; 28: 197-211.

[30] Schepers GWH. The mineral content of the lung in chronic berylliosis. *Dis. Chest.* 1962; 42: 600-607.

[31] Saltini C, Winestock K, Kirby M, Pinkston P, Crystal RG. Maintenance of alveolitis in patients with chronic beryllium disease by beryllium-specific helper T cells. *N. Engl. J. Med.* 1989; 320: 1103-1109.

[32] Falta MT, Pinilla C, Mack DG, Tinega AN, Crawford F, Giulianotti M, Santos R, Clayton GM, Wang Y, Zhang X, Maier LA, Marrack P, Kappler JW, Fontenot AP. Identification of beryllium *J. Exp. Med.* 2013; 210:1403-1418.

[33] Fontenot AP, Kotzin BL, Comment CE, Newman LS. Expansions of T-cell subsets expressing particular T-cell receptor variable regions in chronic beryllium disease. *Am. J. Respir. Cell Mol. Biol.* 1998; 18: 581-589.

[34] Fontenot AP, Falta MT, Freed BM, Newman LS, Kotzin BL. Identification of pathogenic T cells in patients with beryllium-induced lung disease. *J. Immunol.* 1999; 163: 1019-1026.

[35] Chain JL, Martin AK, Mack DG, Maier LA, Palmer BE, Fontenot AP. Impaired function of CTLA-4 in the lungs of patients with chronic beryllium *J. Immunol.* 2013; 191:1648-1656.

[36] Bowerman NA, Falta MT, Mack DG, Wehrmann F, Crawford F, Mroz MM, Maier LA, Kappler JW, Fontenot AP. Identification of multiple public TCR J. Immunol. 2014 May 15;192(10):4571-80. doi: 10.4049/jimmunol.1400007. *Epub.* 2014 Apr 9.

[37] Maier LA, McGrath DS, Sato H, Lympany P, Welsh K, Du Bois R, Silveira L, Fontenot AP, Sawyer RT, Wilcox E, Newman LS. Influence of MHC class II in susceptibility to beryllium sensitization and chronic beryllium disease. *J. Immunol.* 2003; 171: 6910–6918.

[38] McCanlies EC, Ensey JS, Schuler CR, Kreiss K, Weston A. The association between HLA-DPB1Glu69 and chronic beryllium disease and beryllium sensitization. *Am. J. Ind. Med.* 2004; 46: 95–103.

[39] Richeldi L, Sorrentino R, Saltini C. HLA-DPB1 glutamate 69: a genetic marker of beryllium disease. *Science* 1993; 262: 242-244.

[40] Rossman MD, Stubbs J, Lee CW, Argyris E, Magira E, Monos D. Human leukocyte antigen class II amino acid epitopes: susceptibility and progression markers for beryllium hypersensitivity. *Am. J. Respir. Crit. Care Med.* 2002; 165: 788–794.

[41] Saltini C, Richeldi L, Losi M, Amicosante M, Voorter C, van den Berg-Loonen E, Dweik RA, Wiedemann HP, Deubner DC, Tinelli C. Major histocompatibility locus genetic markers of beryllium sensitization and disease. *Eur. Respir. J.* 2001; 18: 677–684.

[42] Wang Z, White PS, Petrovic M, Tatum OL, Newman LS, Maier LA, Marrone BL. Differential susceptibilities to chronic beryllium disease contributed by different Glu69 HLA-DPB1 and -DPA1 alleles. *J. Immunol.* 1999; 163: 1647–1653.

[43] Silveira LJ, McCanlies EC, Fingerlin TE, Van Dyke MV, Mroz MM, Strand M, Fontenot AP, Bowerman N, Dabelea DM, Schuler CR, Weston A, Maier LA. Chronic beryllium disease, HLA-DPB1, and the DP peptide binding groove. *J. Immunol.* 2012; 189: 4014-4023.

[44] Amicosante M, Berretta F, Rossman M, Butler RH, Rogliani P, van den Berg-Loonen E, Saltini C. Identification of HLA-DRPheβ47 as the susceptibility marker of hypersensitivity to beryllium in individuals lacking the berylliosis-associated supratypic marker HLA-DPGluβ69. *Respir. Res.* 2005; 6: 94.

[45] Gaede KI, Amicosante M, Schurmann M, Fireman E, Saltini C, Müller-Quernheim J. Function associated transforming growth factor-beta gene polymorphism in chronic beryllium disease. *J. Mol. Med.* 2005; 83: 397-405.

[46] Richeldi L, Kreiss K, Mroz MM, Zhen B, Tartoni P, Saltini C. Interaction of genetic and exposure factors in the prevalence of berylliosis. *Am. J. Ind. Med.* 1997; 32: 337–340.

[47] Yucesoy B, Johnson VJ. Genetic variability in susceptibility to occupational respiratory sensitization. *J. Allergy* (Cairo). 2011;2011: 346719. doi: 10.1155/2011/346719. Epub 2011 Jun 12.

[48] Newman LS, Kreiss K, King TE Jr, Seay S, Campbell PA. Pathological and immunological alterations in early stages of beryllium disease. Re-examination of disease definition and natural history. *Am. Rev. Respir. Dis.* 1989; 139: 1479-1486.

[49] Seidler A, Euler U, Müller-Quernheim J, Gaede KI, Latza U, Groneberg D, Letzel S. Systematic review: progression of beryllium sensitization to chronic beryllium disease. *Occup. Med.* (Lond) 2012; 62: 506-513.

[50] Newman LS, Buschman DL, Newell JD Jr, Lynch DA. Beryllium disease: assessment with CT. *Radiology* 1994; 190: 835-840.

[51] Freiman DG, Hardy HL. Beryllium disease. The relation of pulmonary pathology to clinical course and prognosis based on a study of 130 cases from the U.S. beryllium case registry. *Hum. Pathol.* 1970; 1: 25-44.

[52] Newman LS. Metals that cause sarcoidosis. *Semin. Respir. Infect.* 1998; 13: 212-220.

[53] Newman LS, Rose CS, Maier LA. Sarcoidosis. *N. Engl. J. Med.* 1997; 336: 1224–1234.

[54] Newman LS, Orton R, Kreiss K. Serum angiotensin converting enzyme activity in chronic beryllium disease. *Am. Rev. Respir. Dis*. 1992; 146: 39-42.

[55] Newman LS, Kreiss K. Nonoccupational beryllium disease masquerading as sarcoidosis: identification by blood lymphocyte proliferation response to beryllium. *Am. Rev. Respir. Dis.* 1992; 145: 1212–1214.

[56] Rossman MD, Kreider ME. Is chronic beryllium disease sarcoidosis of known etiology? *Sarcoidosis Vasc. Diffuse Lung Dis.* 2003; 20: 104-109.

[57] Yoshida T, Ohnuma A, Horiuchi H, Harada T. Pulmonary fibrosis in response to environmental cues and molecular targets involved in its pathogenesis. *J. Toxicol. Pathol.* 2011; 24: 9-24.

[58] Curtis GH. The diagnosis of beryllium disease, with special reference to the patch test. *AMA Arch. Indust. Health* 1959; 19: 150-153.

[59] Schreiber J, Zissel G, Greinert U, Galle J, Schulz KH, Schlaak M, Müller-Quernheim J. Diagnostik der chronischen Berylliose. *Pneumologie* 1999; 53; 193-198.

[60] Stange AW, Furman FJ, Hilmas DE. The beryllium lymphocyte proliferation test: relevant issues in beryllium health surveillance. *Am. J. Ind. Med.* 2004; 46: 453-462.

[61] Rossman MD, Kern JA, Elias JA, Cullen MR, Epstein PE, Preuss OP, Markham TN, Daniele RP. Proliferative response of bronchoalveolar lymphocytes to beryllium. A test for chronic beryllium disease. *Ann. Intern. Med.* 1988; 108: 687-693.

[62] Meyer KC. Beryllium and lung disease. *Chest* 1994; 106: 942-946.

[63] Mroz MM, Kreiss K, Lezotte DC, Campbell PA, Newman LS. Reexamination of the blood lymphocyte transformation test in the diagnosis of chronic beryllium disease. *J. Allergy Clin. Immunol.* 1991; 88: 54-60.

[64] Stokes RF, Rossman MD. Blood cell proliferation response to beryllium: analysis by receiver-operating characteristics. *J. Occup. Med.* 1991; 33: 23-28.

[65] Deubner DC, Goodman M, lannuzzi J. Variability, predictive value, and uses of the beryllium blood lymphocyte proliferation test (BLPT): preliminary analysis of the ongoing workforce survey. *Appl. Occup. Environ. Hyg.* 2001; 16: 521-526.

[66] Ribeiro M, Fritscher LG, Al-Musaed AM, Balter MS, Hoffstein V, Mazer BD, Maier LA, Liss GM, Tarlo SM. Search for chronic beryllium

disease among sarcoidosis patients in Ontario, Canada. *Lung* 2011; 189: 233-241.

[67] Müller-Quernheim J, Gaede KI, Fireman E, Zissel G. Diagnosis of chronic beryllium disease within cohorts of sarcoidosis patients. *Eur. Respir. J.* 2006; 27: 1190-1195.

[68] Welch LS, Ringen K, Dement J, Bingham E, Quinn P, Shorter J, Fisher M. Beryllium disease among construction *Am. J. Ind. Med.* 2013; 56: 1125-1136.

[69] Mikulski MA, Sanderson WT, Leonard SA, Lourens S, Field RW, Sprince NL, Fuortes LJ. Prevalence of beryllium sensitization among Department of Defense conventional munitions workers at low risk for exposure. *J. Occup. Environ. Med.* 2011; 53: 258-265.

[70] Mikulski MA, Leonard SA, Sanderson WT, Hartley PG, Sprince NL, Fuortes LJ. Risk of beryllium sensitization in a low-exposed former nuclear weapons cohort from the Cold War era. *Am. J. Ind. Med.* 2011; 54:194-204.

[71] Van Dyke MV, Martyny JW, Mroz MM, Silveira LJ, Strand M, Cragle DL, Tankersley WG, Wells SM, Newman LS, Maier LA. Exposure and genetics increase risk of beryllium sensitisation and chronic beryllium disease in the nuclear weapons industry. *Occup. Environ. Med.* 2011; 68: 842-848.

[72] Van Dyke MV, Martyny JW, Mroz MM, Silveira LJ, Strand M, Fingerlin TE, Sato H, Newman LS, Maier LA. Risk of chronic beryllium disease by HLA-DPB1 E69 genotype and beryllium exposure in nuclear workers. *Am. J. Respir. Crit. Care Med.* 2011; 183: 1680-1688.

[73] Entzian P, Pawelek W, Lindner B, Schlaak M, Zabel P, Müller-Quernheim J. Grenzen des Berylliumnachweises mit der Laser-Mikrobereichs-Massenspektrometrie (LAMMS). *Pneumologie* 1998; 52: 674-679.

[74] Verma DK, Ritchie AC, Shaw ML. Measurement of beryllium in lung tissue of a chronic beryllium disease case and cases with sarcoidosis. *Occup. Med.* 2003; 53: 223-227.

[75] Apostoli P, Schaller KH. Urinary beryllium-a suitable tool for assessing occupational and environmental beryllium exposure? *Int. Arch. Occup. Environ. Health* 2001; 74: 162-166.

[76] Wegner R, Heinrich-Ramm R, Nowak D, Olma K, Poschadel B, Szadkowski D. Lung function, biological monitoring, and biological effect monitoring of gemstone cutters exposed to beryls. *Occup. Environ. Med.* 2000; 57: 133-139.

[77] Müller-Quernheim J. Chronic beryllium disease. Orphanet encyclopedia, November 2005. Http://www.orpha.net/data/patho/GB/uk-CBD.pdf.

[78] Müller-Quernheim J, Zissel G, Schopf R, Vollmer E, Schlaak M. Differentialdiagnose Berylliose/Sarkoidose bei einem Zahntechniker. *Dtsch. Med. Wochenschr.* 1996; 121: 1462-1466.

[79] Sood A, Beckett WS, Cullen MR. Variable response to long-term corticosteroid therapy in chronic beryllium disease. *Chest* 2004; 126: 2000-2007.

[80] Baughman RP, Lower EE. Steroid-sparing alternative treatments for sarcoidosis. *Clin. Chest Med.* 1997; 18: 853-864.

[81] Müller-Quernheim J, Kienast K, Held M, Pfeifer S, Costabel U. Treatment of chronic sarcoidosis with an azathioprine/prednisolone regimen. *Eur. Respir. J.* 1999; 14: 1117-1122.

[82] Salvator H, Gille T, Hervé A, Bron C, Lamberto C, Valeyre D. Chronic beryllium disease: azathioprine as a possible alternative to corticosteroid treatment. *Eur. Respir. J.* 2013; 41: 234-236.

[83] Schweisfurth H. Infliximab zur Therapie der Sarkoidose. *Atemw-Lungenkrkh.* 2011; 37: 12-18.

[84] Maier LA, Barkes BQ, Mroz M, Rossman MD, Barnard J, Gillespie M, Martin A, Mack DG, Silveira L, Sawyer RT, Newman LS, Fontenot AP. Infliximab therapy modulates an antigen-specific immune response in chronic beryllium disease. *Respir. Med.* 2012; 106: 1810-1813.

[85] Sood A. Current treatment of chronic beryllium disease. *J. Occup. Environ. Hyg.* 2009; 6: 762-765.

[86] U.S. Environmental Protection Agency, IRIS Database for Risk Assessment. Toxicological review of beryllium and compounds (CAS No. 7440-41-7). April 1998. http://www.epa.gov/iris/.

[87] Kelleher PC, Martyny JW, Mroz MM, Maier LA, Ruttenber AJ, Young DA, Newman LS. Beryllium particulate exposure and disease relations in a beryllium machining plant. *J. Occup. Environ. Med.* 2001; 43: 238-249.

[88] Kriebel D, Brain JD, Sprince NL, Kazemi H. The pulmonary toxicity of beryllium. *Am. J. Respir. Crit. Care Med.* 1988; 137: 464-473.

[89] Sprince NL, Kanarek DJ, Weber AL, Chamberlin RI, Kazemi H. Reversible respiratory disease in beryllium workers. *Am. Rev. Respir. Dis. 1978*; 117: 1011-1017.

[90] Kriebel D, Sprince NL, Eisen EA, Greaves IA. Pulmonary function in beryllium workers: assessment of exposure. *Br. J. Ind. Med.* 1988; 45: 83-92.

[91] Bartell SM, Ponce RA, Takaro TK, Zerbe RO, Omenn GS, Faustman EM. Risk estimation and value-of-information analysis for three proposed genetic screening programs for chronic beryllium disease prevention. *Risk Anal.* 2000; 20: 87-99.

[92] Thomas CA, Deubner DC, Stanton ML, Kreiss K, Schuler CR. Long-term efficacy of a program to prevent beryllium disease. *Am. J. Ind. Med.* 2013; 56: 733-741.

[93] Schweisfurth H. Berylliose. *Atemw-Lungenkrkh.* 2012; 38: 227-234.

In: Beryllium
Editor: Pauleen Dyer

ISBN: 978-1-63321-590-0

Chapter 3

COSMOGENIC NUCLIDE BERYLLIUM-10 IN MARGINAL SEA RESEARCH

Yong-Liang Yang[1*], Masashi Kusakabe[2], Zhenxia Liu[3], Chengde Shen[4], Tiegang Li[5], Xuefa Shi[3] and Zhenbo Cheng[3]

[1]National Research Center of Geoanalysis, Beijing, China
[2]Nakaminato Laboratory, National Institute of Radiological Sciences, Hitachinaka, Japan
[3]First Institute of Oceanography, State Oceanic Administration, Qingdao, China
[4]Guangzhou Institute of Geochemistry, Chinese Academy of Sciences, Guangzhou, China
[5]Institute of Oceanology, Chinese Academy of Sciences, Qingdao, China

ABSTRACT

The application of cosmogenic nuclide beryllium-10 (^{10}Be) in marginal sea study has been discussed. Dissolved ^{10}Be concentration profiles in seawaters of the East China Sea and the Kuroshio Current have been investigated. The results show that ^{10}Be concentrations in this area are mainly controlled by surface biological productivity, particle remineralization, and the degree of mixing with the Yangtze River and the Kuroshio waters. Generally the ^{10}Be water depth profiles can be divided into three layers: the surface mixing layer, the particulate ^{10}Be regeneration layer and the bottom layer. Vertical distributions of ^{10}Be show that ^{10}Be is enriched in the bottom waters near the Yangtze River estuary and the central continental shelf. Box model results indicate that ^{10}Be input from the Kuroshio Current is more important than the Yangtze River input and atmospheric precipitation. About 81% of the ^{10}Be input to the East China Sea is scavenged into the sediments and 19% of the ^{10}Be flows out of the East China Sea by currents and water exchange. The ^{10}Be sedimentation flux in the East China Sea is nearly five times higher than the average global ^{10}Be production rate. Therefore the East China Sea may be an important sink for ^{10}Be. ^{10}Be records in two sediment cores from the Okinawa Trough during the last glacial period and the Holocene are discussed. The average ^{10}Be sedimentation flux during the last glacial period was more than three times higher than that at present and also higher than the ^{10}Be accumulation fluxes in the Pacific open ocean during the glacial period, indicating the conveyor rule of the Kuroshio Current and a boundary scavenging effect for ^{10}Be. The high ^{10}Be concentrations and fluxes in the Okinawa Trough during the glacial period proved that on millennium scale the Kuroshio Current still flowed in the Okinawa Trough area with significantly high intensity. The ^{10}Be minimum during the Younger Dryas cold event may indicate that the Kuroshio had been once greatly weakened or even terminated in the Okinawa Trough area which could be related to the response of the Pacific Ocean to the Younger Dryas.

Keywords: Beryllium-10, sediment core, East China Sea, Kuroshio, last glacial maximum

INTRODUCTION

Beryllium-10, a cosmogenic radionuclide with half-life of 1.5×10^6 yr (McHargue and Damon, 1991), is produced primarily in the atmosphere by spallation of ^{14}N or ^{16}O with cosmic rays through nuclear reactions such as ^{14}N(n, p α)^{10}Be and ^{16}O(p, x)^{10}Be. Once produced in the atmosphere, ^{10}Be is removed within a year from the atmosphere by wet and dry precipitation (Heikkila et al., 2009). ^{10}Be has been widely used as a radiometric dating tool. It has been used for estimation of exposure age of meteorites, loess, boulders on moraines from the last glacial due to glacial successions and bedrock landslides due to glacial erosion. In marine geology, it has been used for chronology of deep-sea sediment cores, Mn-Fe nodules and crusts (von Blanckenburg et al., 1996; Christl et al., 2003; Cheng et al., 2006). ^{10}Be is an excellent tracer in paleo-climatological and paleo-oceanographic research.

Recently, due to the success of ice core collections in the Polar areas and the Himalaya Mountains, ^{10}Be in ice cores has been investigated for reconstruction of changes in geomagnetism and atmospheric water transport in the past. In the open oceans where particulate matters in surface water are less abundant, ^{10}Be behaves like typical nutrients. Its concentration increases with the water depth and the age of deep waters (Kusakabe et al., 1987; Ku et al., 1990). As the sedimentation flux of ^{10}Be is influenced by deposition flux of particulate materials in the water column, ^{10}Be can be used as a geochemical tracer for particle scavenging in the water column and a proxy for paleo-productivity (Anderson et al., 1990; Frank et al., 1995).

In the past decades, many studies on application of ^{10}Be in marine environment have been carried out on paleo-environmental events associated with changes in ocean currents (e.g. Lao et al., 1992; Muscheler et al., 2000). Studies on applications of cosmogenic nuclide ^{10}Be in deep-sea environment have achieved great successes. However, these studies were mainly focused on the deep ocean and abyssal sediments, few studies have been on the coastal and marginal seas. In this chapter, applications of ^{10}Be as a tracer of ocean currents in a marginal sea---the East China Sea, are introduced. ^{10}Be input to the sea is through dry and wet precipitation as well as river runoff. The East China Sea is one of the largest marginal seas in the northwest Pacific with a broad continental shelf and high river runoff from the Yangtze River. In summer the river discharge reaches a maximum of 45,000 m^3 s^{-1}, with a sediment load of 4.68×10^8 t yr^{-1} (Beardsley et al., 1985). Due to the rich supply of nutrients, the East China Sea is an area of high productivity and particle scavenging of trace elements. The Kuroshio Current flows over the continental slope and mixes with the shelf water. Therefore, the East China Sea can be regarded as a mixing area for the ^{10}Be from the Yangtze River and the ^{10}Be from the open ocean carried by the Kuroshio Current.

Changes in ^{10}Be concentrations in sediments may be caused by climatic factors (ocean currents, ocean export productivity, terrigenous material dilution, etc.), or change in atmospheric ^{10}Be in productivity (changes in solar activity and geomagnetic field) ^{10}Be changes in sediments can be more complex than in ice cores. The signals of the effects of paleo-climatic events on the marginal sea areas, especially the East China Sea which is directly affected by ocean circulation, may have been amplified in the sediment records. The aim of this chapter is (1) to summarize the studies on application of ^{10}Be in marginal seas (the East China Sea and the Yellow Sea); (2) to gain an insight into the influence of the water circulation on ^{10}Be distribution and budget in the water column and sediments in the East China Sea; and (3) the role of marginal seas in the global cycle of ^{10}Be. Understanding of these processes will shed light on the oceanographical response of marginal seas to climate change, and the fate of particle-reactive elements in continental shelf areas.

HYDROLOGICAL SETTING OF THE RESEARCH AREA

The continental shelf of the East China Sea has a total area of about 7.5 $\times 10^5$ km^2; one of the largest in the world. It is also one of the most productive areas of the world oceans. The major water masses in the East China Sea are Yangtze River Diluted Water, Continental Coastal Water, the shelf water, and the Kuroshio branch (Miao et al., 1987). Due to the shallow water depth, seasonal variation of water circulation in the East China Sea is very large (Yanagi et al., 1996). In spring and summer the surface water is transported seaward, and the bottom water moves landward as compensation. In winter, northeasterly winds drive the surface water toward the land, while the bottom water is transported seaward due to Ekman transport and compensation. A numerical model by Yanagi et al. (1996) shows that autumn water circulation is similar to the winter circulation rather than the summer circulation.

The Kuroshio Current (Figure 1) is a thin narrow band less than l00 km in width and about l km at maximum depth running for 3000 km along the western edge of the Pacific between the Philippines and the east coast of Japan. It flows northeastward along the eastern margin of the continental shelf and the Okinawa Trough. Sun (1987) demonstrated that the Kuroshio axis was close to the East China Sea continental shelf margin in fall and winter, and shifted offshore in spring and summer. Lin et al. (1992) analyzed year-long satellite IR images and found that the Kuroshio front extended onto the shelf in winter and retreated to the shelf edge in summer and fall. Kuroshio water has been divided into the Kuroshio Surface Water (KSW), Kuroshio Subsurface Water (KSSW), and Kuroshio Intermediate Water (KIW) (Chen, 1995). In the southern East China Sea off northeast of Taiwan, the impingement of the Kuroshio onto the continental shelf induces upwelling of the Kuroshio Subsurface Water (Gong et al., 1996), and the Kuroshio Intermediate Water (Chen, 1995). This upwelled Kuroshio branch flows northward and mixes with the shelf water.

The Kuroshio is one of the world's major ocean currents, and plays a vital role in the circulation of the North Pacific. Its variations can have disastrous effects on climate of the North Pacific. Many studies have been carried out to gain more knowledge about the western boundary currents in general, the Kuroshio in particular, and the way in which climate responds to changes in conditions at sea.

The Okinawa Trough is a seabed feature of the East China Sea. It is an active, initial back-arc rifting basin which has formed behind the Ryukyu arc-trench system in the West Pacific. It developed where the Philippine Sea Plate is subducting under the Eurasia Plate. It has a large section more than 1,000 meters (3,300 ft) deep and a maximum depth of 2,716 meters. The thick sediment layers on the Okinawa Trough sea bed provide abundant high-resolution information on paleoenvironment and paleoceanography in the North Pacific.

METHODS

Sample Collection

(1) Sea waters

Seawater samples were collected from the Japanese R/V *Kaiyo* during two cruises in October 1993 (K93-05) and in August 1994 (K94-04) (Figure 2). Water samples from 12 stations on the main observation line (PN line) from the north of Okinawa (71°30′ N, 128° 15′ E) to the Yangtze River mouth (31° 15′ N, 123° 00′ E) and one station C-2 (K94-04 cruise only) were collected with the General Oceanic Rosette Multi-bottle Array System (20L×24 bottles). Immediately after sampling water samples were filtered through a filter with a pore size of 0.4 mm and acidified with concentrated HCl to pH 2–3. Then, Beryllium and Fe carriers were added, and ^{10}Be was co-precipitated with ferric hydroxide by addition of ammonia.

(2) Surface sediments

Surface sediment samples were collected from the East China Sea (station 478-B and 553-B), the southwest mud area in the Yellow Sea (seven stations), and the northern Okinawa Trough (two stations) (Table 1 and Figure 1).

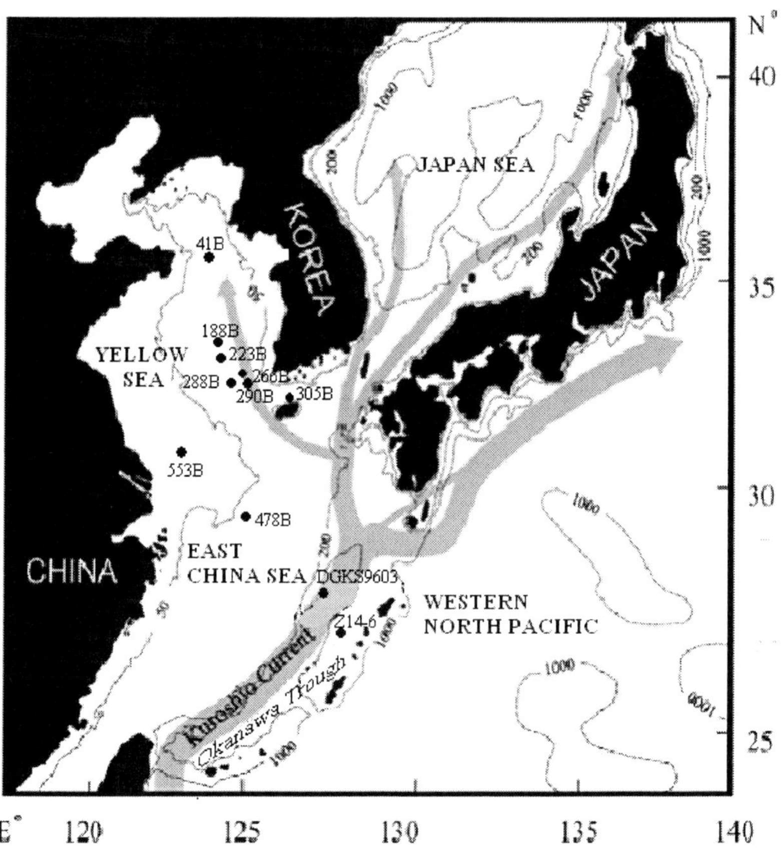

Figure 1. Locations of Okinawa Trough and sediment sampling sites. The shaded arrow denotes the Kuroshio and its branches; Numbers in isobars indicate water depth in meter.

(3) Sediment cores

Two piston cores (DGKS9603 and Z14-6) were collected from the Okinawa Trough (Figure 1). The gravity core DGKS9603 (hereinafter referred to as core 03 in the following discussions) (28°08.869′N, 127°16.238′E, water depth: 1,100 m, length: 5.85 m) was collected from the middle section of the north Okinawa Trough during the joint cruise of French R/V *L'Atalante* in 1996 by the State Oceanic Administration of China and the French Development Research Institute of Maritime. The core was dominated by silt clay, containing large amount of foraminifera and several layers of foraminifera sand, apparently without turbidities.

The gravity core Z14-6 were taken in the northeast of the Okinawa Trough (27° 07′ N, 127° 27′ E, water depth: 739 m, core length: 896 cm) during a Chinese research vessel *Science No. 1* cruise in 1993 by Institute of Oenology, Chinese Academy of Science. The sediments of the entire core were clay and silt clay layer containing a large number of foraminifera and foraminifera sand, no significant turbidities were observed.

Extraction and Analysis

Extraction

For seawater samples, beryllium targets were prepared for AMS (accelerator mass spectrometer) analyses of ^{10}Be from the hydroxide precipitation using the procedure of Kusakabe et al. (1987). Briefly, ^{10}Be was purified by using anion and cation exchange chromatography, DIBK (di-isobutyl ketone) solvent extraction, precipitation of $BeOH_2$; and conversion to BeO by heating to 1000 °C in a Pt crucible.

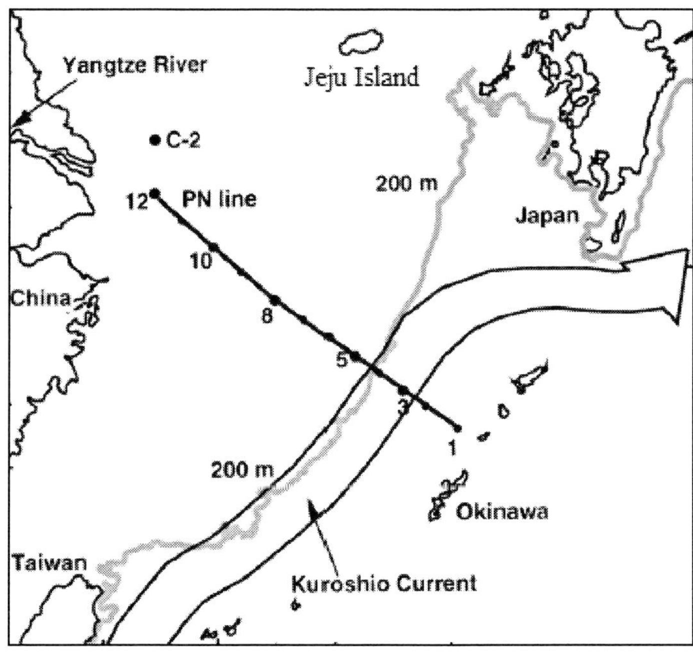

Figure 2. Water sampling sites in the East China Sea.

Table 1. Sediment sampling locations

Station	Longititude	Latitude	Water Depth (m)	Sampling period
DGKS9603	127° 16.238'E	28° 08.869'N	1100	1996
Z14-6	127°27´E	27°07´N	739	1990
41-B	123°29.904'E	35°47.876'N	73	1998.9.33
188-B	124°19.703'E	33°35.408'N	73	1998.10.7
223-B	124°20.261'E	33°11.603'N	70	1998.10.9
288-B	124°30.105'E	32°35.497'N	49	1998.10.11
290-B	125°10.327'E	32°35.960'N	67	1998.10.11
266-B	125°04.845'E	32°47.720'N	70	1998.10.10
305-B	126°25.295'E	32°23.655'N	105	1998.10.10
478-B	125°05.754'E	29°36.138'N	80	1998.10.20
553-B	122°59.862'E	30°59.880'N	49	1998.10.15

Total of 47 sediment samples for ^{10}Be concentrations were prepared in Guangzhou Institute of Geochemistry, Chinese Academy of Sciences using the following procedures (Shen et al., 2004). One gram of dried sample was leached using 6 M HCl for overnight. An aliquot was used for ^{9}Be and trace metal analysis. About 1.0 mg of Be carrier was added to the ^{10}Be aliquot. Ammonia was added to form co-precipitation of $Be(OH)_2$ and $Al(OH)_3$. Further purification was made using a cation exchange column (Dowex 50W-X8) and an anion exchange column (Dowex 1W-X8). The eluted Be fractions were precipitated as $Be(OH)_2$ by ammonia water. The precipitates were then dried in small quartz beakers and placed in a gold-inner-coated electric furnace and temperatures were ramped to 1,000 °C to convert the material to BeO.

AMS Analysis

^{10}Be in seawater samples was measured at the AMS facility of Lawrence Livermore National Laboratory, USA. They were mixed with a four-fold amount of Ag powder and were pressed into 1 mm diameter holes of the AMS sample holders. The AMS measurements of ^{10}Be in sediment samples were performed on 5MeV accelerator mass spectrometer 5UD Pelletron (NEC, National Electrostatics Corporation, USA) at Micro Analysis Laboratory, Tandem Accelerator (MALT), the University of Tokyo. The standard samples for AMS were prepared from the ^{10}Be stock solution (4.98×10^{-10}) originally supplied by ICN Co., which had already been proven to produce a very good agreement in the ^{10}Be/^{9}Be ratio with the NIST standard. The reproducibility in

$^{10}Be/^{9}Be$ ratio of several standard samples through repeated measurements was 2%~4% (1σ).

Be and Pb analysis

Concentrations of beryllium (Be) and lead (Pb) were obtained by inductively coupled plasma-mass spectrometry (ICP–MS) at the Guangzhou Institute of Geochemistry, Chinese Academy of Sciences (Yi et al., 1996).

RESULTS AND DISCUSSIONS

Berylium-10 As a Tracer of the Kuroshio Current

(1) ^{10}Be concentration in the East China Sea water columns

The horizontal distribution of dissolved ^{10}Be concentration in surface waters and vertical profiles of ^{10}Be concentration from the East China Sea and the Okinawa Trough are shown in Figure 3 and Figure 4, respectively. The errors are based on 1σ uncertainties from the AMS measurements.

Several characteristics can be seen from these data. Firstly, ^{10}Be concentrations (60–320 atoms g^{-1} seawater) in surface waters in the continental shelf part of the East China Sea (PN-5~PN-12) are much lower than the surface water ^{10}Be concentration (600–800 atoms g^{-1} seawater) in the Pacific Ocean (Kusakabe et al., 1987), suggesting intensive particle scavenging of ^{10}Be in this ocean margin. Secondly, ^{10}Be concentration in the surface water increases sharply at the Okinawa Trough stations PN-2, PN-3 and PN-4, reflecting the influence of the Kuroshio Current on ^{10}Be concentration in water columns. Thirdly, generally there are surface mixing layers in which ^{10}Be concentrations are homogeneous and the thickness of the mixing layers increases towards the Okinawa Trough. The ^{10}Be concentrations at station PN-3 in the Kuroshio Current are nearly homogeneous in the upper 200 m water because the thermocline in the Kuroshio Current may reach as deep as 200 m. Below the mixing layer, ^{10}Be concentrations generally increase with water depth, reflecting the regeneration of ^{10}Be scavenged from the overlying surface waters due to biological particles degradation. ^{10}Be enrichment was observed in the bottom water at two sites in the summer cruise (Figure 4), one near the Yangtze River estuary where biological productivity is high and the other in the central continental shelf (PN-7 and PN-8). Finally, seasonal changes of ^{10}Be profiles were observed at some stations with lower ^{10}Be

concentrations in 1994 summer than in 1993 autumn. PN-10 in summer had lower ^{10}Be concentrations compared with the adjacent stations.

Based on the above-mentioned features of ^{10}Be profiles in the East China Sea, three layers in the water column can be identified. A surface layer represents the mixed layer in which productivity is high and ^{10}Be is depleted. The second layer is a depth interval in which ^{10}Be regeneration occurs due to the particulate matter degradation. The contour plots (Figure 4) further depicts that ^{10}Be was enriched (relative to the Kuroshio water) in the bottom waters near the Yangtze River estuary and in the central continental shelf. Biogenic particles and terrestrial particles both influence the distribution of ^{10}Be in this area. However, along the PN line, sediments on the seabed are mainly coarse-sized silts (Saito and Yang, 1993). Beryllium has stronger affinity to clay-sized particles than to silt-sized particles (Olsen et al., 1986). Thus, the ^{10}Be sorption on the coarse terrestrial particles at the PN stations should be negligible. The high scavenging of ^{10}Be by particles in the surface water transports ^{10}Be into the subsurface water where ^{10}Be is subjected to regeneration with the remineralization of biogenic particles. Since the coarse terrestrial sediments cannot scavenge ^{10}Be effectively, the regenerated ^{10}Be either should be enriched in the bottom water or be carried elsewhere by water circulation. The contour plots show that the ^{10}Be is enriched (relative to the Kuroshio water) in bottom waters near the Yangtze River estuary and in the central continental shelf. It should be noted that this enrichment of dissolved ^{10}Be in the bottom water may occur only in the stratified waters. We did not investigate the ^{10}Be distribution in winter season when the stratification disappears. From hydrographic data it can be expected that in winter some of the ^{10}Be will be transported to the open ocean by bottom currents and the Kuroshio (Yanagi et al., 1996).

(2) ^{10}Be concentration in the Kuroshio water

^{10}Be concentrations in the Kuroshio Current water (stations PN-2 and PN-3) are generally high (700–800 atoms g^{-1} seawater), comparable to the open-ocean surface water values (500– 1000 atoms g^{-1} seawater, Raisbeck et al., 1979, 1980; Kusakabe et al., 1987) in the Pacific Ocean. Since the Kuroshio originates from the tropical open ocean area of the Pacific with low nutrients and low productivity, it is expected that the scavenging of ^{10}Be in the Kuroshio surface water should be low. Another reason that the higher ^{10}Be concentration in the Kuroshio water may lie in the fact that it is formed in the tropical ocean (off the east of Philippines) where rain precipitation is high, hence leading to a higher ^{10}Be input to seawater (Somayajulu et al., 1984).

It has been suggested that the Kuroshio Intermediate Water, which is rich in nutrients compared to the surface water, is the major source of nutrients to the East China Sea (Chen et al., 1995; Chen, 1996). ^{10}Be has long residence time in the deep waters (600–1000 years, Raisbeck et al., 1980; Kusakabe et al., 1987) and behaves like a nutrient. Therefore it is evident that the Kuroshio is a significant ^{10}Be contributor to the East China Sea.

(3) The mean residence times of ^{10}Be

The surface mixed layers in the East China Sea range from 20 to 80 m thick. The concentrations of ^{10}Be in the mixed layers and the whole water column in the continental shelf part of the East China Sea range from 60- 200 atoms g^{-1} and 60-700 atoms g^{-1}, respectively. If we use the global mean ^{10}Be production rate of (1:2±0:26) $\times 10^6$ atoms cm^{-2} yr^{-1} (Monaghan et al., 1985), we can roughly estimate the ^{10}Be residence time in the surface layers and in the whole water column of the East China Sea. The estimated mean residence times in the surface waters range from 9 h near the Yangtze River estuary to 27 days near the Okinawa Trough. The mean residence time for ^{10}Be in the surface Kuroshio was estimated to be 47 days.

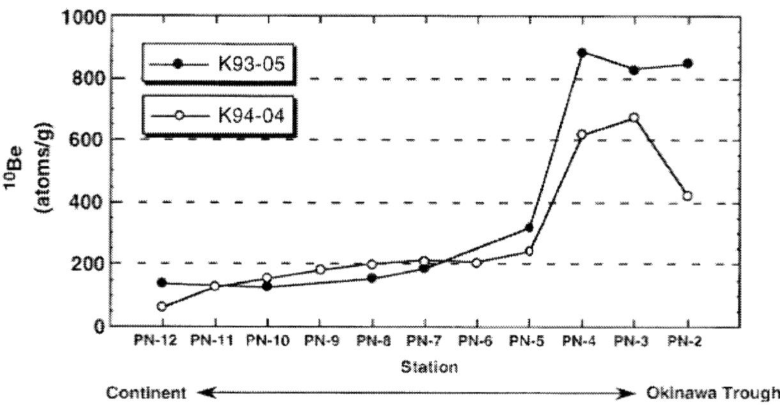

Figure 3. Horizontal distribution of ^{10}Be concentration in surface waters along the PN line (Yang et al., 2003).

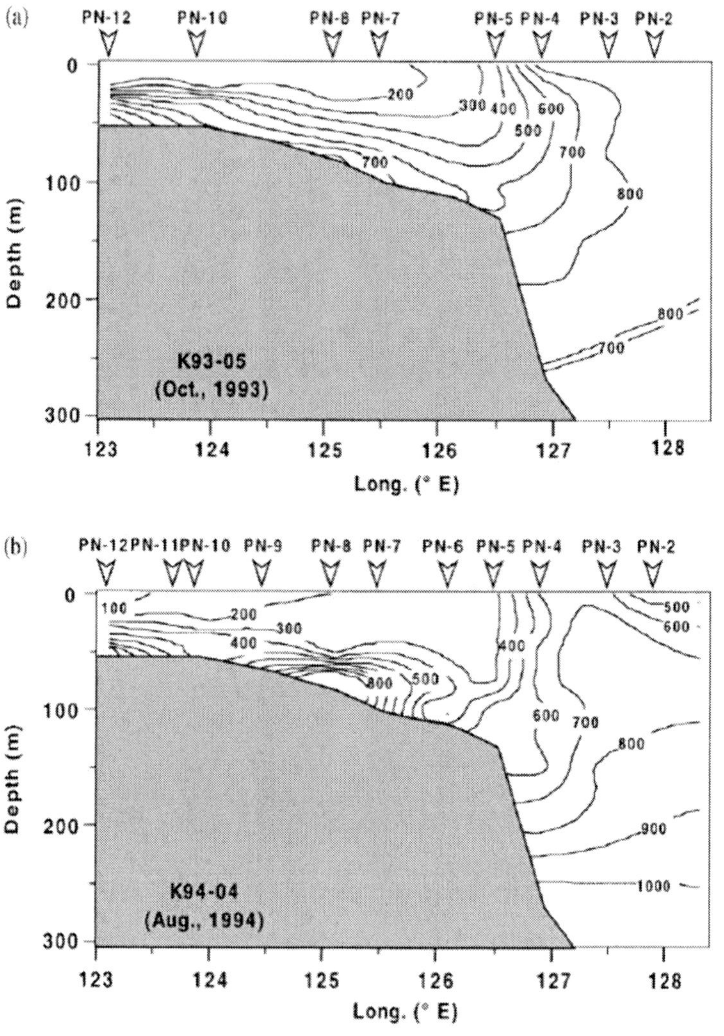

Figure 4. ^{10}Be concentration (atoms g^{-1}) contour plots on PN line in (a) 1993 autumn and (b) 1994 summer. Arrows indicate sampling stations. (Yang et al., 2003).

These estimations do not take into account of lateral advection and diffusion supply of ^{10}Be; and therefore should be viewed as an upper limit of the ^{10}Be residence times in the East China Sea and the Okinawa Trough. In addition, since the atmospheric flux of ^{10}Be to the area that is controlled by the ^{10}Be production rate in the atmosphere and the annual rainfall (Somayajulu et al., 1984) should not be the same as the global mean ^{10}Be production rate, the residence time derived from the above calculation may include significant amount of error. Yet, compared with the ^{10}Be residence time (0.5 yr) in the surface mixed layer of the San Nicolas Basin in the eastern Pacific Ocean margin (Raisbeck et al., 1980; Kusakabe et al., 1982), the scavenging residence time of ^{10}Be is much shorter in the East China Sea.

(4) The ^{10}Be budget in the East China Sea

In order to understand further the role of the Kuroshio intrusion, it is necessary to have a ^{10}Be budget in the East China Sea. To accomplish this, the following simple box model was employed. The balance of ^{10}Be in the East China Sea can be expressed by the equation

$$Q_R C_R + Q_K C_K + I_a A t = Q_S C_S + S \quad (1)$$

where Q_R and Q_K are the water fluxes (km^3/6 months) of the Yangtze River, and the Kuroshio for the 6 month wet season into the East China Sea, respectively; Q_S is the water flux for the 6 month wet season flowing out of the East China Sea. The data of Q_R (813 km^3/6 months), Q_K (27,360 km^3/6 months), and Q_S (28,593 km^3/6 months) are adopted from Kim (1992), Yanagi (1994), Chen (1996), and Q_K is the sum of the Kuroshio surface, subsurface and intermediate water fluxes (Chen, 1996). I_a is the atmospheric input of ^{10}Be; which is assumed to be equivalent to the average global ^{10}Be production rate $(1.2\pm0.26)\times10^6$ atoms cm^{-2} yr^{-1}; Monaghan et al., 1985), A is the area of the East China Sea (7.5×10^5 km^2/6 months); C_R, C_K, and C_S are ^{10}Be concentrations in the Yangtze River water, the Kuroshio water, and the shelf water, respectively. S is the sedimentation flux of ^{10}Be buried in sediments. Here, we neglect the contribution from the waters coming through the Taiwan Straight and assume that ^{10}Be concentration in the Yangtze River water is the same as the average dissolved ^{10}Be concentrations in several North American rivers and the Pearl River, China (3220±1960 atoms g^{-1} water, Kusakabe et al., 1991). The results show that the ^{10}Be flux into the East China Sea from the river, the Kuroshio, and the atmosphere are $(2.6\pm1.6)\times10^{21}$, 1.9×10^{22}, and $(4.5\pm1.0)\times10^{21}$ atoms/6 months; respectively. Therefore, the atmospheric contribution to the ^{10}Be concentration in the East China Sea is comparable to that from the Yangtze River but the Kuroshio ^{10}Be contribution is an order of magnitude higher.

The total ^{10}Be flux into the East China Sea is $(2.6\pm0.2)\times10^{22}$ atoms/6 months, which should balance the ^{10}Be flux flowing out of the East China Sea and the ^{10}Be flux to sediments. Next, we can calculate the flowing flux of ^{10}Be out of the East China Sea. We use the surface ^{10}Be concentrations (average 172±1.5 atoms g^{-1} seawater for the stations on PN line except the two stations PN-3 and PN-4 located in the Kuroshio Current) as the ^{10}Be concentration in the shelf water that flows out of the East China Sea. The reason for the choice of surface ^{10}Be as C_S is that only the surface water in summer in the East China Sea can flow out due to the water circulation pattern, i.e. the surface water flows seaward and the bottom water flows landward. The estimated outflow flux is $(4.9\pm0.04)\times10^{21}$ atoms/6 months; and the sedimentation flux is $(2.1\pm0.2)\times10^{22}$ atoms/6 months; which is 81% of the total ^{10}Be input and equivalent to $(5.6\pm0.5)\times10^{6}$ atoms cm^{-2} yr^{-1}, almost five times of the average global ^{10}Be production rate (1.2×10^{6} atoms cm^{-2} yr^{-1}). Taking into account of the possible contribution by the water from the Taiwan Strait that we neglected in our model, the ^{10}Be sedimentation flux should be even higher, so the value of 5.6×10^{6} atoms cm^{-2} yr^{-1} is only a lower limit. Taking into account of the errors in the above calculations, the ^{10}Be sedimentation fluxes in the East China Sea are 3–6 times higher than the average global ^{10}Be production rate. Therefore, the East China Sea is an important ^{10}Be sink. Since the sand and silt sized sediments are not expected to scavenge ^{10}Be efficiently, most of the ^{10}Be sedimentation may be focused in the clay-sized sediment areas north of the PN line and in the near-shore areas. This needs to be proved by further work. It should be noted that the above conclusion is only valid for the spring and summer seasons. In the autumn and winter, the water circulation pattern (also the surface biogenic productivity) will change, and therefore the proportions of the ^{10}Be flowing-out flux and sedimentation flux in the total ^{10}Be output flux should also be changed.

About 19% of the ^{10}Be input to the East China Sea will eventually flow out of the East China Sea. The outlets for the East China Sea and the Yellow Sea waters are limited in summer season, as the Kuroshio Current effectively blocks the shelf water flowing to the open Pacific Ocean due to the density difference. The only available outlets are the Taiwan Straits (to the South China Sea) and mixing with the Tsushima Current (to the Japan Sea)(Nozaki et al., 1989), and both are very shallow. From the above calculation, it is clear that ^{10}Be in the East China Sea is effectively trapped in summer by the continental shelf.

In summary, the Kuroshio plays a role of a conveyor belt, which carries ^{10}Be from the open ocean into the East China Sea via mixing with the shelf water. ^{10}Be can therefore be used as a tracer of the Kuroshio water. ^{10}Be record in sediment cores collected from the Okinawa Trough should reflect the ancient Kuroshio evolution during the last glacial period. Understanding of these processes will help to further depict the Kuroshio behavior and changes in the Late Pleistocene and the Holocene, and this will be discussed in the following section.

Historical Record of ^{10}Be in Sediment Cores from the Okinawa Trough

(1) The last glacial maximum and the cold events in Late Pleistocene and Holocene

The world's most recent glacial period or "ice age" began about 110,000 years ago and ended around 12,500 years ago. The maximum extent of this glacial period was the Last Glacial Maximum (LGM) that occurred around 18,000 years ago when ice sheets were at their maximum extension (Elenga et al., 2000), marking the peak of the last glacial period. During this time, vast ice sheets covered much of North America, northern Europe and Asia. These ice sheets profoundly impacted Earth's climate, causing drought, desertification, and a dramatic drop in sea levels.

The transition from the cold Pleistocene to the warm Holocene occurred about 10,000 years ago. The Pleistocene/Holocene boundary is placed at 11,600 calendar years BP (cal. yr before present). This abrupt warming was preceded by the ca. 1300-yr-long cold reversal, called the Younger Dryas (YD). The Younger Dryas was first recognized in European pollen sequences in 1901 (Hartz and Milthers, 1901). The Younger Dryas is seen most clearly in Greenland ice cores (Johnsen et al., 1992; Alley et al., 1993). Various proxies indicate contemporaneous climate changes at more southerly latitudes, e.g. in southern China (Wang et al., 2001), but these changes were not everywhere of the same severity, or in the same direction.

Heinrich events, first described by marine geologist Hartmut Heinrich, occurred during the last glacial period. During such events, armadas of icebergs broke off from glaciers and traversed the North Atlantic. Six distinct events in marine sediment cores of mud retrieved from the sea floor can be distinguished, which are labeled H1~H6 going back in time. There is some evidence that H3 and H6 differ from other events. The icebergs' melting caused prodigious amounts of fresh water to be added to the North Atlantic. Such inputs of cold, fresh water may have altered the density-driven thermohaline circulation patterns of the ocean, and often coincide with indications of global climate fluctuations.

Currently there are mainly two different views on the issue whether or not the Kuroshio flowed through the Okinawa Trough during the last glacial period. Some researchers have hypothesized that the Kuroshio migrated out of the Okinawa Trough area during the last glacial period possibly caused by a land-bridge which formed when the sea level lowered in the East China Sea continental shelf during the last glacial period, connecting Taiwan and the Ryukyu islands (Ujiié and Ujiié, 1999) and preventing the Kuroshio from entering the Okinawa Trough area. Others hold that the Kuroshio still flowed in the Okinawa Trough area at the last glacial maximum but with different intensity and meander paths (Xu and Oda, 1999; Li et al., 2001; Lan et al., 2003). The debates so far are based on the $\delta^{18}O$ records, occurring frequency in preferential species of planktonic foraminifers (Shieh and Chen, 1995; Jian et al., 2000), and grain-size distributions of sediments (Wang, 1990; 1999) in the Okinawa Trough, which can reflect variations of the Kuroshio Current in different respects. In order to better understand the history of the Kuroshio Current, use of beryllium isotopic tracer as an alternative indicator is needed to solve the issue. In this section, ^{10}Be records in the Okinawa Trough sediment cores since the last glacial period are discussed and comparison with the ^{10}Be records in ice cores from polar areas is made.

(2) Chronology of the sediment cores

The measurement of $\delta^{18}O$ for sediment core DGKS9603 was carried out using *Globigerinoids sacculifer*, and 240 measured values of $\delta^{18}O$ were corrected with the PDB standard value with error of ± 0.08‰). Fourteen AMS ^{14}C ages (Table 2) were determined on *Globorotalia menardii*, *Globigerinoids sacculifer* or *Neogloboquadrina dutertrei* at Bate Analyses Co., USA. Corrected ages were obtained after subtraction of the difference of 400 yr between atmosphere and seawater. Ages younger than 20 kyr BP were calibrated to the international calendar age using CALIB 4.0 (Liu et al., 2000) and ages older than 20 kyr BP using the method of Laj et al. (1996). Accumulation fluxes of ^{10}Be were calculated based on the calendar ages and the dry bulk density of sediments. Liu et al., (2000) discussed ancient climate change in core DGKS9603 based on the foraminifera transfer function analysis using the internationally accepted calendar age.

The ^{14}C ages obtained from AMS measurements for cores DGKS9603 and Z14-6 are shown in Table 3 and Table 4, respectively. According to the oxygen isotope chronology, Yan and Cang (1990) found that the bottom of Core Z14-6 reached the marine oxygen isotope stage 6 (MIS-6)(Yan and

Cang, 1990). In the following discussion, the time scale established by Yan and Cang (1990) is used for core Z14-6 (Table 4).

Table 2. AMS Radiocarbon dating of core DGKS9603 (From Liu et al., 2000)

Core depth (cm)	Foraminifera species	Corrected age (yr BP)*	Calendar age (yr BP)	Sedimentation rate (cm kyr^{-1})
11~13	*Globorotalia menardii*	2740	2840	4.38
26~31	*Globorotalia menardii*	4500	5300	9.4
47~49	*Globigerinoides sacculifer*	8080	8950	5.4
63~65	*Globigerinoides sacculifer*	9690	11090	9.9
89~94	*Globigerinoides sacculifer*	11230	13140	17.0
128~130	*Globigerinoides sacculifer*	12980	15510	21.4
174~179	*Globigerinoides sacculifer*	16580	19640	13.2
209~211	*Globigerinoides sacculifer*	19070	22510	13.5
300~302	*Neogloboquadrina dutertrei*	26350	29790	12.5
368~373	*Globigerinoides sacculifer*	27960	31400	43.2
420~424	*Globigerinoides sacculifer*	32300	35110	11.9
448~450	*Neogloboquadrina dutertrei*	37410	39900	5.3
484~486	*Neogloboquadrina dutertrei*	39020	41400	22.4
530~532	*Neogloboquadrina dutertrei*	41260	42800	20.5

*^{13}C corrected age.

(3) Historical record of ^{10}Be in sediment cores from the Okinawa Trough

I. Core DGKS9603

The ^{10}Be concentrations in core DGKS9603 are shown in Table 3. The variations of ^{10}Be concentrations and fluxes together with the foraminiferal $\delta^{18}O$ and SST in core DGKS9603 were plotted with age in Figure 5. The errors of ^{10}Be concentrations and fluxes are based on 1σ uncertainty in the AMS analysis. The concentrations of ^{10}Be range from 2.9×10^8 to 9.1×10^8 atoms g^{-1} with the minimum at the Younger Dryas event. The fluxes of ^{10}Be range from about 6.2×10^8 to 5.0×10^9 atoms cm^{-2} kyr^{-1} with the maximum occurring at 23,000 yr BP and the minimum at the Younger Dryas event. The average ^{10}Be flux in the Holocene was 1.12×10^9 atoms cm^{-2} kyr^{-1}, nearly the same as the present atmospheric ^{10}Be production rate [$(1.21\pm0.26) \times 10^9$ atoms cm^{-2} kyr^{-1}](Monaghan et al., 1985), whereas the average ^{10}Be flux during the height of the last glacial period (22,000~15,000 yr BP) was 4.53×10^9 atoms cm^{-2} kyr^{-1}, three times higher than the present atmospheric ^{10}Be production rate, indicating the conveyor rule of the Kuroshio Current and a boundary scavenging effect.

The ^{10}Be concentrations did not show obvious decrease at the Heinrich events H1~ H3, but was obviously low at H4. The most striking feature in the ^{10}Be curve is the abrupt changes in both ^{10}Be concentration and flux during the Younger Dryas (YD) event. The Younger Dryas event was reflected in core DGKS9603 between 12.2~11.2 cal. kyr BP with a peak at 11.4 cal. kyr BP which seems to have a 600 year lag compared with the YD event occurred in the GISP2 ice core from Summit in Central Greenland (Finkel and Nishiizumi, 1997). The SST showed a clear peak at the YD.

Several factors may cause the difference between the ^{10}Be fluxes in the last glacial period and the Holocene observed in core DGKS9603. First, the high ^{10}Be flux in the Okinawa Trough during the last glacial period should be correlated with a weakened geomagnetic intensity that induced a higher global ^{10}Be production rate (Christl et al., 2003). Beryllium-10 production rate is proportional to the flux of cosmic rays, which is modulated by solar activity and the strength of the Earth's magnetic field (Christl et al., 2003). Weakening of the magnetic field allows more cosmic rays to impinge on the Earth's atmosphere, thereby increasing ^{10}Be production. Lao et al. (1992) reported that the ocean-wide average accumulation rate of ^{10}Be in Pacific sediments, which may reflect the global average production rate of ^{10}Be, was at least 25% greater during the height of the last glacial period (24,000−16,000 yr BP) than during the Holocene. They proposed that the higher production rate of ^{10}Be be caused by the lower intensity of the geomagnetic field during that period.

Table 3. Analytical results for Be isotopes in core DGKS9603

Age (kyr BP)	^{9}Be ($\mu g \cdot g^{-1}$)	^{10}Be $\times 10^{8}$ atoms$\cdot g^{-1}$	^{10}Be sedimentation flux $\times 10^{9}$atom$\cdot cm^{-2}$ ka^{-1}
0.69	0.47	7.18 ± 0.01	0.93 ± 0.01
1.60	0.47	6.71 ± 0.11	0.87 ± 0.01
3.20	0.52	6.48 ± 0.11	0.84 ± 0.01
3.71	0.45	6.04 ± 0.09	0.78 ± 0.01
4.86	0.43	5.96 ± 0.10	1.63 ± 0.03
6.45	0.39	5.29 ± 0.09	1.44 ± 0.02
8.48	0.45	5.98 ± 0.09	0.95 ± 0.01
9.70	0.43	6.98 ± 0.09	1.47 ± 0.02
11.70	0.69	2.92 ± 0.37	0.62 ± 0.08
12.23	0.74	7.49 ± 0.09	2.09 ± 0.03

Age (kyr BP)	^{9}Be (μg· g^{-1})	^{10}Be ×10^{8} atoms·g^{-1}	^{10}Be sedimentation flux ×10^{9}atom· cm^{-2} ka^{-1}
12.70	0.73	7.36 ± 0.12	2.05 ± 0.03
13.14	0.67	7.51 ± 0.13	2.83 ± 0.05
13.42	0.67	8.47 ± 0.10	4.04 ± 0.05
13.50	0.74	8.13 ± 0.13	3.88 ± 0.06
13.59	0.75	7.09 ± 0.09	3.38 ± 0.04
14.13	0.77	8.04 ± 0.13	3.83 ± 0.06
16.59	0.69	8.05 ± 0.11	3.84 ± 0.05
16.72	0.75	7.35 ± 0.12	3.51 ± 0.06
17.96	0.78	7.60 ± 0.13	4.39 ± 0.07
18.18	0.81	7.39 ± 0.14	4.27 ± 0.08
20.65	0.71	7.64 ± 0.09	4.41 ± 0.05
21.87	0.79	8.71 ± 0.15	5.03 ± 0.08
23.26	0.78	8.60 ± 0.14	3.14 ± 0.05
23.48	0.67	8.42 ± 0.10	3.07 ± 0.04
26.58	0.70	8.86 ± 0.24	3.30 ± 0.09
26.80	0.76	8.30 ± 0.56	3.10 ± 0.21
27.83	0.69	7.75 ± 0.09	2.89 ± 0.04
29.30	0.73	8.95 ± 0.11	3.34 ± 0.04
30.48	0.67	7.79 ± 0.10	2.62 ± 0.03
31.23	0.70	8.71 ± 0.10	2.94 ± 0.03
31.96	0.68	9.12 ± 0.11	3.07 ± 0.04
36.19	0.78	5.48 ± 0.81	0.79 ± 0.12
42.64	0.67	6.81 ± 0.09	3.86 ± 0.05
46.43	0.40	7.62 ± 0.08	4.31 ± 0.04

Sea-level changes may have caused the low ^{10}Be flux in the Okinawa Trough during the Holocene. At the end of the last glacial, the sea-level rise in the East China Sea enlarged the sedimentation area, which dispersed the ^{10}Be brought by the Kuroshio Current, leading to significantly reduction in ^{10}Be flux in the Okinawa Trough. The ^{10}Be inventory in the glacial sediment during 12,000–22,000 yr BP was four times greater than that in the Holocene (0–10,000 yr BP). If the glacial ^{10}Be were evenly distributed in the East China Sea which is three times larger in area than the Okinawa Trough area and the same ^{10}Be concentration in sediment is assumed, there would be still at least 25% more ^{10}Be inventory in the glacial sediment.

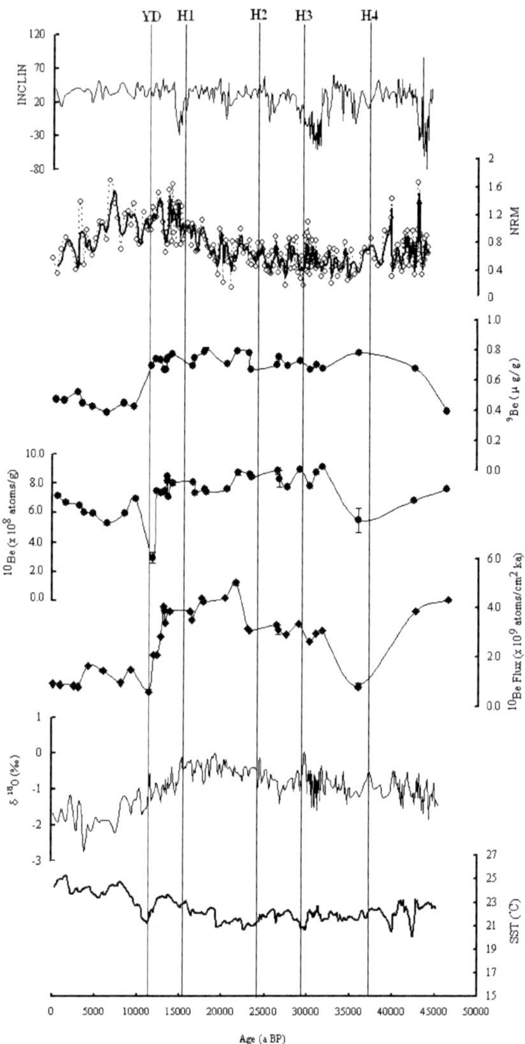

Figure 5. Comparison of ^{10}Be flux and ^{10}Be, ^{9}Be concentrations and proxy climate data (δ^{18}O and winter surface seawater temperature, SST), NRM (natural remnant magnetism with unit of 10^{-8} Tesla), and INCLIN (inclination with unit of degree) versus age for core DOKS9603 since 47,000 yr BP. The Younger Dryas (YD) cold climate event was reflected in core DGKS9603 between 12.2~11.2 cal. kyr BP with a peak at 11.4 cal. kyr BP. The vertical lines from left to right denote the timings of the Younger Dryas and the Heinrich events H1, H2, H3, and H4, respectively.

The higher ^{10}Be concentration and flux in the last glacial may also have been caused by higher biological production in the Okinawa Trough with a stronger boundary scavenging during that period, leading to a lateral transport of ^{10}Be to the Okinawa Trough and higher ^{10}Be accumulation in sediments as occurred in the eastern equatorial Pacific and the sub-Antarctic ocean and Weddell Sea (Anderson et al., 1990; Frank et al., 1995). This kind of lateral transport of ^{10}Be is generally assisted by advection. ^{10}Be may have been transported by the North Equatorial Current (NEC) from the Pacific pool, followed by northward transport of the Kuroshio to the Okinawa Trough area. If a land-bridge had formed between Taiwan and Ryukyu Islands (Ujiié and Ujiié, 1990), such a lateral transport would not have happened by eddy diffusion alone. Therefore, it can be excluded that large-scale focused ^{10}Be sedimentation with the flux three-times greater than that in the average Pacific Ocean could happen in the Okinawa Trough area without the Kuroshio Current entering the Okinawa Trough.

The possibility may be excluded that the enhanced ^{10}Be flux during the last glacial period was caused by direct input of the Yangtze River water into the Okinawa Trough due to extended river channel on the exposed East China Sea continental shelf when the sea-level descended during the last glacial period. Based on the mass balance of ^{10}Be in the East China Sea, the Yangtze River contribution to ^{10}Be in seawater of the East China Sea is only one-sixths of that from the Kuroshio Current. If the amount of ^{10}Be brought by the Yangtze River during the last glacial period is assumed to be equal to that in the Holocene, i.e. $(2.6\pm1.6)\times10^{21}$ atoms yr^{-1} and all deposited in the Okinawa Trough, the accumulation of ^{10}Be due to this source would have been 1.18×10^{9} atoms cm^{-2} kyr^{-1}, which is about one-fourths of the observed ^{10}Be accumulation flux during the last glacial period. Therefore it can be concluded that ^{10}Be in the Okinawa Trough during the last glacial period was mainly brought from the Kuroshio which remained flowing in the Okinawa Trough area with significantly high intensity on millennial scale. The lower intensity of the geomagnetic field during the glacial period lead to more ^{10}Be production in the atmosphere and more ^{10}Be input to the Pacific Ocean pool, which in turn increased amount of ^{10}Be carried by the Kuroshio Current into the Okinawa Trough area.

The variation in ^{10}Be flux in the Holocene may also reflect the fluctuations of the Kuroshio Current during that period. Based on the records of the preferential planktonic foraminiferal species for the Kuroshio in two sediment cores from the southern and northern Okinawa Trough, Lan et al. (2003) pointed out that the Kuroshio may have narrowed and weakened twice at 7 kyr~8 kyr BP and 1 kyr~4 kyr BP and proposed that the main axis of the Kuroshio once moved outside the Okinawa Trough during the last glacial period and re-entered into the Okinawa Trough during the minimum event of *P. obliquiloculata* at 6,500 yr BP. Ujiié & Ujiié (2000) also identified an minimum event of *Pulleniatina,* a tropical characteristic species of the Kuroshio, during 4,500-3,000 yr BP in the Okinawa Trough and suggested that the event may be the evidence of the re-entering of the Kuroshio into the Okinawa Trough. Our ^{10}Be results show that the ^{10}Be accumulation flux in the Okinawa Trough was reduced at about 3,000 and 8,000 yr BP, suggesting possible Kuroshio weakening during the two events.

The observation of long Be residence time in the central open Pacific Ocean (ca. 1,000 yr)(Raisbeck et al., 1980) suggests that the Pacific Ocean could be a reservoir for both ^{10}Be and ^{9}Be. Beryllium-9 concentrations in core DGKS9603 was also significantly higher in the last glacial period than those in the Holocene that could be a result of a higher input of terrestrial material into the Okinawa Trough (Xiong and Liu, 2004). However, the ^{9}Be was probably brought by the Kuroshio Current from the Pacific open ocean. The ^{9}Be concentrations also showed a significant response during the last deglaciation, reflecting the sea-level rise in the East China Sea at the end of the glacial period.

The ^{10}Be minimum in core DGKS9603 during the Younger Dryas event seems to reflect a global ^{10}Be event which has been reported for ice cores and lacustrine sediment cores (Finkel and Nishiizumi, 1997). The lack of correlation between ^{10}Be flux and climate proxy $\delta\ ^{18}$O in Summit ice core GISP2 suggests that ^{10}Be was not affected by any change of climate during its transport and deposition to the Summit, and ^{10}Be flux almost did not show any change even during the rapid increase period of $\delta\ ^{18}$O at the of Younger Dryas/Holocene transition (Finkel and Nishiizumi, 1997). However, the ^{10}Be minimum event in the Okinawa Trough during the Younger Dryas seems to relate both to a change in production rate and an abrupt change in ocean during that time, which caused the Kuroshio flowed less into the Okinawa Trough area.

The difference between the ^{10}Be flux variations in sediment from the Okinawa Trough and in the Greenland ice cores lies in that ^{10}Be records in the Okinawa Trough had been subject to much larger influence from climate change factors such as current circulations and sea-level changes, whereas ^{10}Be records in the Greenland ice cores reflect more directly the change in the ^{10}Be production. The ^{10}Be change in the Summit ice core GISP2 during 11,500~12,700 yr BP (Finkel and Nishiizumi, 1997) was less than the difference as great as a factor of four in the ^{10}Be fluxes in core DGKS9603 before and after the Younger Dryas, suggesting a magnification of the Younger Dryas signal due to the change in the Kuroshio Current. Therefore, even though the change in the ^{10}Be flux in the Okinawa Trough during the YD was mainly caused by the ^{10}Be production rate, only three-fifth of the observed ^{10}Be flux increase could be accounted for. The more reasonable explanation might be that during the YD the Kuroshio was severely weakened, leading to less ^{10}Be carried by the Kuroshio Current into the Okinawa Trough.

The first evidence could be the abrupt lowering of the sea surface temperature SST during the Younger Dryas, implying that the Kuroshio experienced a weakening or excursion that reduced ^{10}Be input to the Okinawa Trough. Secondly, the ^{10}Be concentrations and flux behaved differently during the YD event. The change in ^{10}Be concentrations only showed a pulse of decrease, and then recovered to the level almost same as the beginning of the pulse, reflecting a response to a climate change signal which could be either the change in the Kuroshio or the sea level change which might cause a large amount terrestrial materials in to the Okinawa Trough, and diluted the ^{10}Be concentrations in sediment. On the other hand, the change in ^{10}Be flux seems to be a quick response to the Younger Dryas event which overlapped a long-term decreasing trend of ^{10}Be flux, probably resulted from an increase of the intensity of geomagnetism as ^{10}Be flux had already shown a decreasing trend even before the YD since 13,000 yr BP. After the YD, the ^{10}Be flux has never recovered to the high level as before. The concentrations and flux of ^{10}Be decreased so low that both reached the minimum levels in the core. The ^{10}Be concentrations was down to 2.9×10^8 atoms g^{-1}, even lower than that in the modern surface sediment in the East China Sea continent shelf (See the following section), and the ^{10}Be flux was even lower than the present average ^{10}Be atmospheric production rate. From this observation, it may be concluded that the change in the Kuroshio was so great during the Younger Dryas event that the Kuroshio either once diverted its direction or even terminated, which might be related to the variations in the North Equatorial Current (NEC) in that period.

In summary, the vertical profile of ^{10}Be in the sediment core presented here have given us a first look at the ^{10}Be distribution in the Okinawa Trough sediment influenced by the geomagnetic intensity and the climate change effect on the Kuroshio Current. The accumulation flux of ^{10}Be in the Okinawa Trough sediment during the last glacial period was much higher than that in the open ocean and than in the Holocene, indicating that the Kuroshio Current was still flowing in the Okinawa Trough area on millennial scale. The ^{10}Be flux was more sensitive to short cooling events such as the Younger Dryas than the ^{10}Be flux records in ice cores, suggesting that in this ocean margin area, the signals of climate variations could be magnified through ^{10}Be records due to the intensity or even switch-on-and-off of the Kuroshio Current. Our results have shown that ^{10}Be can be a promising tracer and proxy in climatic changes in the western boundary of the Pacific Ocean.

II. Core Z14-6

Core Z14-6, located in the middle of the eastern slope of the Okinawa Trough, is currently one of the longest cores collected in the Okinawa Trough with the longest time span. Some detailed studies on planktonic foraminifera, oxygen isotopes, SST, etc. for core Z14-6 have been reported (Yan and Cang, 1990). Based on the geochronology data of core Z14-6, twelve sub-layers covering the Late Pleistocene and Holocene were selected for ^{10}Be analysis. In this section, the vertical distribution of ^{10}Be and ^{9}Be concentrations in core Z14-6, and the mechanism of the changes in ^{10}Be sedimentation flux in the northeastern Okinawa Trough since the Late Pleistocene are discussed, in a attempt to use ^{10}Be as an alternative indicators to determine the behavior and changes of the Kuroshio during the Late Pleistocene and the Holocene. The ^{10}Be records in the northern Okinawa Trough core DGKS9603 are compared with those in core Z14-6.

The measurement results for ^{10}Be and ^{9}Be concentrations and ^{10}Be deposition fluxes at Z14-6 site are listed in Table 4. Figure 6 shows the variations in the ^{10}Be and ^{9}Be concentrations in core Z14-6 with age. The ^{10}Be concentrations in core Z14-6 in general was higher in the Holocene than those in the last glacial period. The average concentration of ^{10}Be was 6.10×10^{8} atoms g^{-1}. The highest value (8.71×10^{8} atoms g^{-1}) was observed at 6.3 kyr BP layers, the lowest value (3.44×10^{8} atoms g^{-1}) at 50 cm depth layer (9.27 kyr BP) in the core. The Holocene ^{10}Be and ^{9}Be appeared high concentrations at 18 and 34 cm layers. Compared with the core 03 in the northern Okinawa Trough, ^{10}Be concentrations in core Z14-6 was generally lower.

The mean ^{10}Be sedimentation flux at Z14-6 was 1.04×10^{9} atoms cm^{-2} kyr^{-1} (Figure 7). The highest value (1.36×10^{9} atoms cm^{-2} kyr^{-1}) occurred at 6.3 kyr BP, and the lowest value (6.45×10^{8} atoms cm^{-2} kyr^{-1}) at 9.27 kyr BP. As the water depth at station Z14-6 is shallow (739 m), and the Kuroshio did not flow through the site Z14-6 during the last glacial period, so the ^{10}Be sedimentation fluxes were only equivalent to the average atmospheric productivity of ^{10}Be (1.21×10^{9} atoms cm^{-2} kyr^{-1})(Monaghan et al., 1985), and it can be concluded that no focusing effect of ^{10}Be deposition at this part of the Okinawa Trough occurred. However, several low ^{10}Be sedimentation fluxes at Z14-6 site may correspond to a number of cold events. The most notable are the Younger Dryas event and Heinrich events H1 and H2. The ^{10}Be fluxes at Z14-6 site decreased around 4 kyr BP and 9 kyr BP and therefore, greater changes in the Kuroshio might presumably occurred in the Holocene. The maximum ^{10}Be sedimentation flux at Z14-6 during the Holocene was significantly higher than in the last glacial ^{10}Be sedimentation flux peak, indicating that even though the water depth at Z14-6 site is shallow, the impact of the weakening of the Kuroshio in the last deglaciation was still significant.

^{9}Be varied basically in synchronization with ^{10}Be in core Z14-6. At the Last Glacial Maximum (LGM), ^{9}Be concentrations were less than those in the Holocene. One possibility is that during the last glacial period high terrigenous materials presented in the eastern Okinawa Trough (Xiong and Liu, 2004). More likely scenario is that a large part of ^{9}Be from the ocean brought by the Kuroshio as discussed for the case of core DGKS9603.

Table 4. Analytical results for Be isotopes in core Z14-6

Core depth (cm)	Age* (kyr BP)	Be concentration (μg g^{-1})	^{10}Be concentration (atoms g^{-1})	^{10}Be sedimentation flux (atoms cm^{-2} kyr^{-1})
2	0.37	0.19	4.68×108	7.32×108
18	3.34	0.37	8.61×108	1.35×109
34	6.30	0.31	8.71×108	1.36×109
50	9.27	0.15	3.44×108	6.45×108
66	12.2	0.27	6.74×108	1.06×109
82	14.1	0.16	4.97×108	8.96×108
98	16.9	0.14	5.63×108	1.02×109
114	19.6	0.19	6.02×108	1.09×109
130	22.4	0.20	4.88×108	8.81×108
146	25.2	0.33	6.67×108	1.20×109
162	27.9	0.39	6.33×108	1.14×109
175	29.3	0.33	6.49×108	1.17×109
Mean		0.25	6.10×108	1.04×109

The Younger Dryas event in the northern Okinawa Trough core 03 occurred at 11.4 cal. kyr BP (^{14}C age of 9.90 kyr BP)(Liu et al., 2000). The decrease of ^{10}Be concentrations in core Z14-6 was 200 yr lag behind the response of the Younger Dryas event in core 03. This phenomenon that responses to the same event differ in different Okinawa Trough areas has been reported. For example, the conversion from the LGM to post-LGM indicated by U^{k}_{37} (a biomarker based on long-chain alkenones) in core DGKS9603 (Meng et al., 2002) was not as clearly as that in core Z14-6 (Zhou et al., 2007), which is interpreted as core DGKS9603 in the western trough was affected by materials from the East China Sea continental slope much larger than at site Z14-6 (Li et al., 2001).

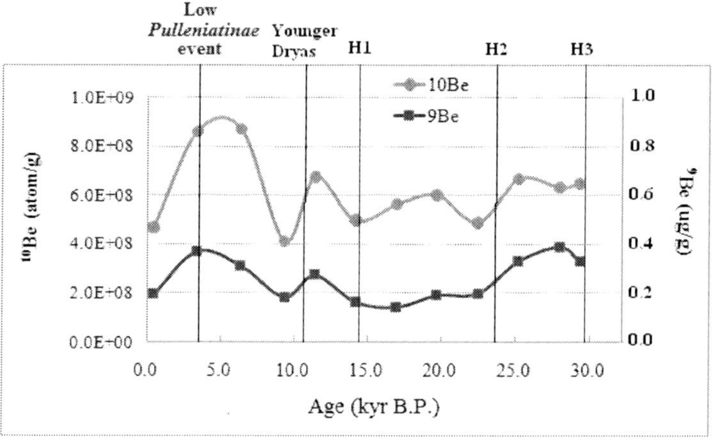

Figure 6. Comparison of ^{10}Be and ^{9}Be concentrations versus age for core Z14-6 since 30,000 yr BP (The vertical lines denote Younger Dryas, Heinrich H1, H2, H3, and low *Pulleniatinae* events).

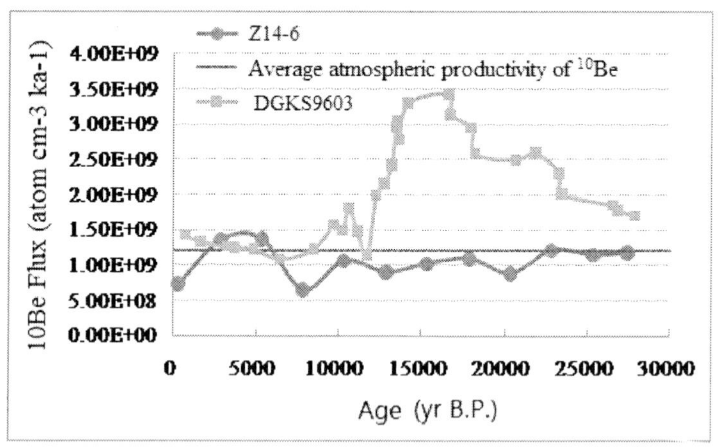

Figure 7. Comparisons of ^{10}Be sedimentation fluxes in Z14-6, DGKS9603 and the global mean ^{10}Be production rate.

Wang et al. (1999) suggested that the ancient Kuroshio might have flowed into the Okinawa Trough by some straits between the Ryukyu Islands during the last glacial maximum. Lan et al. (2003) proposed that the Kuroshio flowed through the Okinawa Trough along the today's 1000 m isobaths during the last glacial maximum, when the width of the Kuroshio was narrow (<50 km). Both views can explain the lower ^{10}Be deposition flux in core Z14-6 compared with that in core 03. Since the water depth at the core Z14-6 is comparatively shallow (739 m), the Kuroshio did not flow through the Z14-6 site during the last glacial, so only the ^{10}Be sedimentation fluxes equivalent to the average atmospheric ^{10}Be productivity is observed. Beside, core Z14-6 is in the eastern part of the Okinawa Trough on the open ocean side (with low impact of the Yangtze River discharge), thus no sedimentation focusing effect on ^{10}Be can be expected. If changes in the geomagnetic field are taken into account, then the average atmospheric ^{10}Be productivity value was likely to differ from the modern values. Therefore, it can not be determined from the ^{10}Be sedimentation fluxes at Z14-6 site only that the Kuroshio was flowing through the Okinawa Trough during the last glacial maximum. In comparison, core DGKS9603 at water depth of 1,100 m, there is ample paleo-microfossil evidence that the site DGKS9603 began to be affected by the Kuroshio during the last glacial period since about 16 kyr BP (Zhou et al., 2007). Notably, the ^{10}Be sedimentation flux peak in core Z14-6 in the Holocene was significantly higher than those in the last glacial period, reaching 1.36×10^9 atoms cm^{-2} kyr^{-1}, and higher than the average modern atmospheric ^{10}Be productivity, indicating that the Kuroshio had then been flowing through Z14-6 site (Figure 7).

A few lower levels of ^{10}Be concentrations in core Z14-6 correspond to some climate events. One of the most notables is the rapid cooling event Younger Dryas and Heinrich events H1 and H2. The most recognized YD events occurred between 11 ~ 10 kyr BP (Broecker et al., 1988). Marked by massive glacial drift in the North Atlantic, the Heinrich events were the rapid cooling event arising from large-scale influx icebergs into the Atlantic. As can be seen from Figure 6 and Figure 7, the ^{10}Be concentrations and sedimentation fluxes at Z14-6 were relatively low in the Heinrich events H1 and H2 in the glacial period. Notably, these responses often occurred after the events, i.e., a time-lag occurred.

In the study of the middle Okinawa Trough, Li et al. (2007) proposed that the strength of the Kuroshio was weak at 15.5 kyr BP, 11.4 kyr BP and 2.8~5.3 kyr BP, and the Kuroshio weakening at 15.5 kyr BP and 11.4 kyr BP correspond to the cold event events of H1 and YD in the North Atlantic.

The influence of the Kuroshio on the Okinawa Trough area began to strengthen at 16 kyr BP, afterwards the Tsushima Current beginning to develop with rising sea levels. At around 8.5 kyr BP, the modern Tsushima formally formed. During this period, the Kuroshio Current experienced two weakening processes.

Between 4,500 yr BP to 300 yr BP, a minimum *Pulleniatina* event occurred in the Okinawa Trough. *Pulleniatina* is considered to be characteristic microfossil species of the Kuroshio. This event has been identified in ten sediment cores from the Okinawa Trough and a core from the northern Ryukyu Trench. Ujiié and Ujiié (1999) presented evidence of the gradual inflow of Kuroshio into the Okinawa Trough. In a comprehensive analysis of the historic records in core DGKS9603 located near the Kuroshio axis, a sediment core CSH1 on the edge of the modern Tsushima warm current and core YSDP102 collected in the mud area of the Yellow Sea, Li et al. (2007) showed that the low *Pulleniatinae* event occurred along with the Kuroshio weakening process, resulting in a deflection of the Tsushima Current mainstream shaft to the Pacific direction around 3 kyr BP, and the influence of the low temperature, high- nutrient substances in cold water mass from the continental shelf on the northern Okinawa Trough was strengthened. Lan et al. (2003) proposed that the narrowing events of the width of the Kuroshio in the Holocene appeared twice at about 7~8 kyr BP and 1~4 kyr BP, with the strength becoming weakened and wavering. Our results showed that ^{10}Be flux weakening at Z14-6 around 4 kyr BP and 9 kyr BP. This observation can be used as the evidence supporting that significant changes in the Kuroshio path occurred twice after the Younger Dryas event.

The impact of the low temperature events on ^{10}Be can be considered from the contribution to ^{10}Be input to the East China Sea and the Okinawa Trough sediments from the Kuroshio. As discussed in the previous sections, the results of ^{10}Be concentrations in seawater and sediment flux in the East China Sea and the Okinawa Trough indicate that the ^{10}Be input contribution from the Kuroshio was almost an order of magnitude higher than the Yangtze River contribution. The Kuroshio plays a role of a conveyor belt, carrying ^{10}Be from the open ocean to the Okinawa Trough area. ^{10}Be in the atmosphere enters the ocean through precipitation. As the Kuroshio is from the Pacific Ocean with much larger area to receive precipitation from the atmosphere than the area of marginal seas, small contribution of ^{10}Be from the shelf water to the Okinawa Trough sediments can be expected. In the events when the Kuroshio became weakened, the impact of shelf water on the Okinawa Trough sediments would be enhanced, resulting in a lower ^{10}Be sedimentation flux in the Okinawa se of beryllium, mainly from the Yangtze River estuary and the coastal eflecting pollution due to human activities. The lead concentration was gh in the center of the South Yellow Sea mud area, suggesting a kind of t of water circulation on Pb levels in the sediments.

ompared to Pb with low water solubility, beryllium as an alkaline earth element with higher water solubility could more easily be transported to t areas. The southern Yellow Sea mud area is up to 400 km away from ore, the sediments were mainly from re-suspension of sediments in the ellow River under-water delta area (about 400 km north of the Yangtze estuary), transported by the Yellow Sea Coastal Current (Guo et al., . The old Yellow River estuarine sediments formed 1,855 years ago, China had not yet started industrialization, therefore show less influence human activities. These sediments are considered to be more "clean" e to the modern Yangtze River sediments. The Okinawa Trough is r away from China mainland and the mud area to the southwest of Jeju , but its lead concentrations was higher than those in the mud area. As kinawa Trough is close to the Ryukyu Islands, there are likely to be ed by human activities.

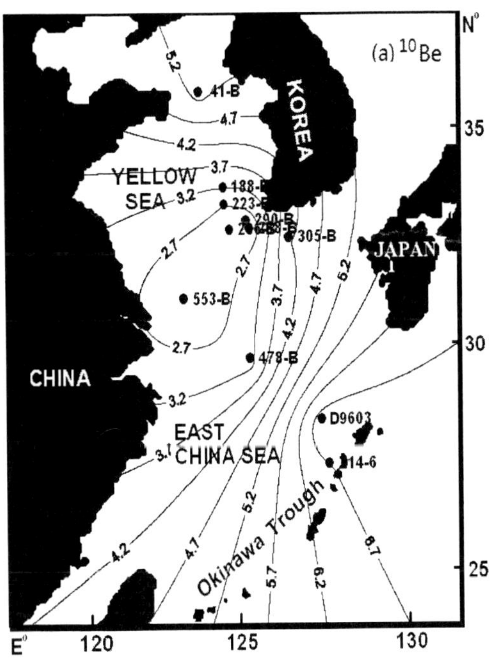

e 8. (Continued)

Trough. Low ^{10}Be level at the Younger Dryas event in the ice cores, lake sediments has been reported (Hu et al., 2003). It is therefore a global phenomenon. Low ^{10}Be event in the northern Okinawa Trough sediments at the Younger Dryas is much more complex than the case of the ^{10}Be change in ice cores and sediments in the Polar Regions. In addition to the short-term drift of the Earth's magnetic field, the low ^{10}Be event in the Okinawa Trough sediments seems to be related with the weakening of the Kuroshio flow caused by sharp change climate, and sharp increase of the proportion of terrigenous sediments. As the result, the ^{10}Be flux at the Younger Dryas event was even below the average modern ^{10}Be atmosphere productivity. In summary, a large change in the Kuroshio in the Younger Dryas event likely occurred.

Spatial Distribution of ^{10}Be in Surface Sediments

(1) Hydrological and sedimentological setting of the East China Sea and Yellow Sea

There are patches of mud sediment areas near the southwest of Jeju Island in the Yellow Sea, and inshore area of the southern East China Sea continental shelf. These mud areas are the accumulation areas for modern fine-grained sediments and pollutants (Park et al., 1992), and also the Yellow Sea and the East China Sea's most important fisheries areas. The central part of the southern Yellow Sea seabed is a shallow plain in the counterclockwise circulation constituted by the flows of the modern Kuroshio branch in the Yellow Sea and the Yellow Sea Warm Current (HWC). Alexander et al. (1991) suggested that materials from the modern Yellow River can reach as far as 32°N, 126°E. Yang et al. (1992) proposed that there is a "cleanest water barrier" between the Yellow Sea continental shelf and deep sea water, and the water barrier is formed by the intrusion of the climbing Kuroshio water along the continental slope. Shen et al. (1993) proposed that the dominant current in the South Yellow Sea shelf is HWC and the Yellow Sea continental shelf sediments can be divided into the cold eddy channel deposition and shelf sediments in the eastern and western sides of HWC. The organic carbon stable isotope composition in South Yellow Sea sediments and suspended matter in seawater indicates that the modern terrigenous sediments in the South Yellow Sea are mainly from the modern Yellow River carrying suspended material (Cai et al., 2001), as well as re- suspended material from the old Yellow River Delta and materials from outside of the Yellow Sea delivered by the Kuroshio. For the South Yellow Sea sediments, the amount and scope of modern terrigenous material from the Yangtze River and Korean Peninsula are limited (Xiong and Liu, 2004).

Table 5. Analytical results for concentrations of ^{9}Be, ^{10}Be, and Pb in the surface sediments

Station	^{9}Be ($\mu g \cdot g^{-1}$)	^{10}Be (10^8 atoms$\cdot g^{-1}$)	Pb ($\mu \cdot g^{-1}$)
DGKS9603	0.47	7.18 ± 0.10	20.6
Z14-6	0.20	6.64 ± 0.05	14.7
41-B	1.03	5.29 ± 0.07	19.8
188-B	0.79	3.11 ± 0.08	15.5
223-B	0.63	2.62 ± 0.05	11.5
266-B	0.63	2.51 ± 0.05	11.3
288-B	1.70	2.45 ± 0.04	15.6
290-B	0.63	2.27 ± 0.04	11.7
305-B	0.56	4.16 ± 0.06	9.9
478-B	0.46	3.09 ± 0.04	10.0
553-B	0.82	2.42 ± 0.04	19.6

In this section, the distribution characteristics of ^{10}Be, ^{9}Be and Pb in 11 surface sediment samples taken from the South Yellow Sea mud area, the East China Sea, and the northern Okinawa Trough are discussed. Beryllium (^{9}Be) and Pb, have completely different sources. Beryllium represents terrigenous material from the river input and Pb is a heavy metal element representing anthropogenic pollution.

(2) ^{10}Be

The analytical results for Be isotopes and Pb in the surface sediments from the South Yellow Sea mud area, the East China Sea, and the northern Okinawa Trough are shown in Table 5 and Figure 8. The analytical errors in ^{10}Be concentrations were based on 1σ uncertainty given by accelerator mass spectrometry. The ^{10}Be concentrations in the surface sediments were in the range of 2.42×10^8~7.18×10^8 atoms g^{-1}, averaging 3.79×10^8 atoms g^{-1}. The ^{10}Be concentrations are in the order of the northern Okinawa Trough > the South Yellow Sea mud area > the East China Sea, indicating that ^{10}Be distribution is mainly affected by the flow of the Kuroshio and its branches, and the contribution to ^{10}Be from the HWC in the Yellow Sea can not be

ignored. The ^{10}Be concentrations in the South Yel
surrounded by the counterclockwise circulation were lov
2.85×10^8 atoms g^{-1}. The station (41-B) at the northern ec
showed an unusually high ^{10}Be concentrations (5.29×1C
8a), suggesting a strong influence by the Kuroshio
concentrations in the two surface sediments from the
Trough (GKS9603 and Z14-6) were comparable, with the
6.64×10^8 atoms g^{-1}, respectively. The ^{10}Be concentrations
East China Sea stations (553-B and 478-B) were 2.42×10^8
g^{-1}, respectively, far below those in the northern Oki
Conclusion can be made that the contribution to ^{10}Be in
Yangtze River water to the East China Sea continental sl
below the Kuroshio's contribution and the high deposition
the East China Sea continental shelf may have a "dilut
concentrations in the sediments.

(3) Beryllium and lead

Figure 8(b) and 8(c) present the distribution patterns (
beryllium (^{9}Be) and lead (Pb) in surface sediments from th
mud area, the East China Sea, and the northern Okinawa T
beryllium concentration in surface sediments from the rese
$\mu g\ g^{-1}$. The beryllium concentration showed a totally dif
that of ^{10}Be, with a decreasing order of the mud area in th
Sea > the East China Sea continental shelf > the north o
with the highest concentration occurring at station 288 (see
concentration in sediments from the South Yellow Sea mu
range of 0.56~1.70 $\mu g\ g^{-1}$, averaging 0.85 $\mu g\ g^{-1}$, The beryl
in surface sediments from the north of Okinawa Trough w
average beryllium concentration in the East China Sea sedin
g^{-1} (the north part) and 0.46 $\mu g\ g^{-1}$ (the east part), respectivel
The Pb concentrations in the sediments from the sou
mud area and the East China Sea continental shelf wer
9.9~20.6 g g^{-1}, averaging 14.6 g g^{-1}, The Pb levels showed a
the northern Okinawa Trough > the East China Sea conti
South Yellow mud area. Except for the northern Okinaw
concentrations in the East China Sea sediments decreased
from the continent. In particular, there is a significant high P
from the Yangtze River estuary to the southern Yellov
characterized by a tongue-shape contour, suggesting that le

the ca
area,
not hi
impac
C
metal
distan
the sh
old Y
Rive
2001)
when
from
relati
furthe
Island
the O
affect

Figur

Trough. Low ^{10}Be level at the Younger Dryas event in the ice cores, lake sediments has been reported (Hu et al., 2003). It is therefore a global phenomenon. Low ^{10}Be event in the northern Okinawa Trough sediments at the Younger Dryas is much more complex than the case of the ^{10}Be change in ice cores and sediments in the Polar Regions. In addition to the short-term drift of the Earth's magnetic field, the low ^{10}Be event in the Okinawa Trough sediments seems to be related with the weakening of the Kuroshio flow caused by sharp change climate, and sharp increase of the proportion of terrigenous sediments. As the result, the ^{10}Be flux at the Younger Dryas event was even below the average modern ^{10}Be atmosphere productivity. In summary, a large change in the Kuroshio in the Younger Dryas event likely occurred.

Spatial Distribution of ^{10}Be in Surface Sediments

(1) Hydrological and sedimentological setting of the East China Sea and Yellow Sea

There are patches of mud sediment areas near the southwest of Jeju Island in the Yellow Sea, and inshore area of the southern East China Sea continental shelf. These mud areas are the accumulation areas for modern fine-grained sediments and pollutants (Park et al., 1992), and also the Yellow Sea and the East China Sea's most important fisheries areas. The central part of the southern Yellow Sea seabed is a shallow plain in the counterclockwise circulation constituted by the flows of the modern Kuroshio branch in the Yellow Sea and the Yellow Sea Warm Current (HWC). Alexander et al. (1991) suggested that materials from the modern Yellow River can reach as far as 32°N, 126°E. Yang et al. (1992) proposed that there is a "cleanest water barrier" between the Yellow Sea continental shelf and deep sea water, and the water barrier is formed by the intrusion of the climbing Kuroshio water along the continental slope. Shen et al. (1993) proposed that the dominant current in the South Yellow Sea shelf is HWC and the Yellow Sea continental shelf sediments can be divided into the cold eddy channel deposition and shelf sediments in the eastern and western sides of HWC. The organic carbon stable isotope composition in South Yellow Sea sediments and suspended matter in seawater indicates that the modern terrigenous sediments in the South Yellow Sea are mainly from the modern Yellow River carrying suspended material (Cai et al., 2001), as well as re- suspended material from the old Yellow River Delta and materials from outside of the Yellow Sea delivered by the Kuroshio. For the South Yellow Sea sediments, the amount and scope of modern terrigenous material from the Yangtze River and Korean Peninsula are limited (Xiong and Liu, 2004).

Table 5. Analytical results for concentrations of ^{9}Be, ^{10}Be, and Pb in the surface sediments

Station	^{9}Be (μg· g^{-1})	^{10}Be (10^{8} atoms·g^{-1})	Pb (μ· g^{-1})
DGKS9603	0.47	7.18 ± 0.10	20.6
Z14-6	0.20	6.64 ± 0.05	14.7
41-B	1.03	5.29 ± 0.07	19.8
188-B	0.79	3.11 ± 0.08	15.5
223-B	0.63	2.62 ± 0.05	11.5
266-B	0.63	2.51 ± 0.05	11.3
288-B	1.70	2.45 ± 0.04	15.6
290-B	0.63	2.27 ± 0.04	11.7
305-B	0.56	4.16 ± 0.06	9.9
478-B	0.46	3.09 ± 0.04	10.0
553-B	0.82	2.42 ± 0.04	19.6

In this section, the distribution characteristics of ^{10}Be, ^{9}Be and Pb in 11 surface sediment samples taken from the South Yellow Sea mud area, the East China Sea, and the northern Okinawa Trough are discussed. Beryllium (^{9}Be) and Pb, have completely different sources. Beryllium represents terrigenous material from the river input and Pb is a heavy metal element representing anthropogenic pollution.

(2) ^{10}Be

The analytical results for Be isotopes and Pb in the surface sediments from the South Yellow Sea mud area, the East China Sea, and the northern Okinawa Trough are shown in Table 5 and Figure 8. The analytical errors in ^{10}Be concentrations were based on 1σ uncertainty given by accelerator mass spectrometry. The ^{10}Be concentrations in the surface sediments were in the range of 2.42×10^{8}~7.18×10^{8} atoms g^{-1}, averaging 3.79×10^{8} atoms g^{-1}. The ^{10}Be concentrations are in the order of the northern Okinawa Trough > the South Yellow Sea mud area > the East China Sea, indicating that ^{10}Be distribution is mainly affected by the flow of the Kuroshio and its branches, and the contribution to ^{10}Be from the HWC in the Yellow Sea can not be ignored. The ^{10}Be concentrations in the South Yellow Sea mud area surrounded by the counterclockwise circulation were low, with an average of 2.85×10^{8} atoms g^{-1}. The station (41-B) at the northern edge of the circulation showed an unusually high ^{10}Be concentrations (5.29×10^{8} atoms g^{-1}) (Figure 8a), suggesting a strong influence by the Kuroshio branch. The ^{10}Be concentrations in the two surface sediments from the northern Okinawa Trough (GKS9603 and Z14-6) were comparable, with the means 7.18×10^{8} and 6.64×10^{8} atoms g^{-1}, respectively. The ^{10}Be concentrations in sediments at two East China Sea stations (553-B and 478-B) were 2.42×10^{8} and 3.09×10^{8} atoms g^{-1}, respectively, far below those in the northern Okinawa Trough area. Conclusion can be made that the contribution to ^{10}Be in sediments from the Yangtze River water to the East China Sea continental shelf sediments is far below the Kuroshio's contribution and the high deposition rate of sediments in the East China Sea continental shelf may have a "dilution" effect on ^{10}Be concentrations in the sediments.

(3) Beryllium and lead

Figure 8(b) and 8(c) present the distribution patterns of concentrations of beryllium (^{9}Be) and lead (Pb) in surface sediments from the South Yellow Sea mud area, the East China Sea, and the northern Okinawa Trough. The average beryllium concentration in surface sediments from the research areas was 0.72 μg g^{-1}. The beryllium concentration showed a totally different pattern from that of ^{10}Be, with a decreasing order of the mud area in the southern Yellow Sea > the East China Sea continental shelf > the north of Okinawa Trough, with the highest concentration occurring at station 288 (see Figure 1). The ^{9}Be concentration in sediments from the South Yellow Sea mud area were in the range of 0.56~1.70 μg g^{-1}, averaging 0.85 μg g^{-1}, The beryllium concentration in surface sediments from the north of Okinawa Trough was 0.34 μg g^{-1}. The average beryllium concentration in the East China Sea sediments were 0.82 μg g^{-1} (the north part) and 0.46 μg g^{-1} (the east part), respectively.

The Pb concentrations in the sediments from the southern Yellow Sea mud area and the East China Sea continental shelf were in the range of 9.9~20.6 g g^{-1}, averaging 14.6 g g^{-1}, The Pb levels showed a decreasing trend: the northern Okinawa Trough > the East China Sea continental shelf > the South Yellow mud area. Except for the northern Okinawa Trough, the Pb concentrations in the East China Sea sediments decreased with the distance from the continent. In particular, there is a significant high Pb level extending from the Yangtze River estuary to the southern Yellow Sea mud area, characterized by a tongue-shape contour, suggesting that lead, different from the case of beryllium, mainly from the Yangtze River estuary and the coastal area, reflecting pollution due to human activities. The lead concentration was not high in the center of the South Yellow Sea mud area, suggesting a kind of impact of water circulation on Pb levels in the sediments.

Compared to Pb with low water solubility, beryllium as an alkaline earth metal element with higher water solubility could more easily be transported to distant areas. The southern Yellow Sea mud area is up to 400 km away from the shore, the sediments were mainly from re-suspension of sediments in the old Yellow River under-water delta area (about 400 km north of the Yangtze Rive estuary), transported by the Yellow Sea Coastal Current (Guo et al., 2001). The old Yellow River estuarine sediments formed 1,855 years ago, when China had not yet started industrialization, therefore show less influence from human activities. These sediments are considered to be more "clean" relative to the modern Yangtze River sediments. The Okinawa Trough is further away from China mainland and the mud area to the southwest of Jeju Island, but its lead concentrations was higher than those in the mud area. As the Okinawa Trough is close to the Ryukyu Islands, there are likely to be affected by human activities.

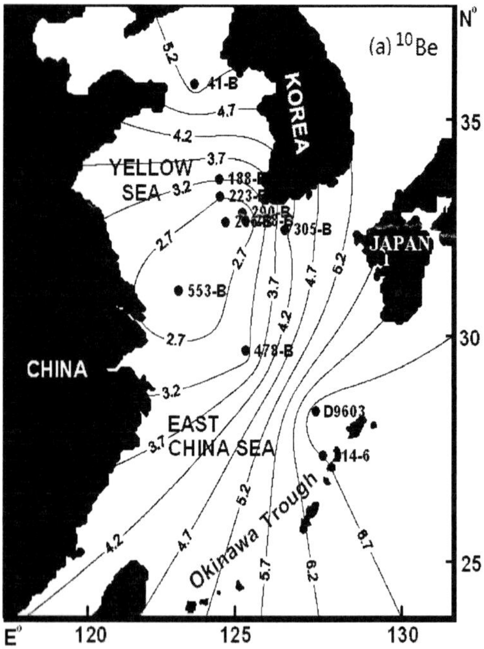

Figure 8. (Continued)

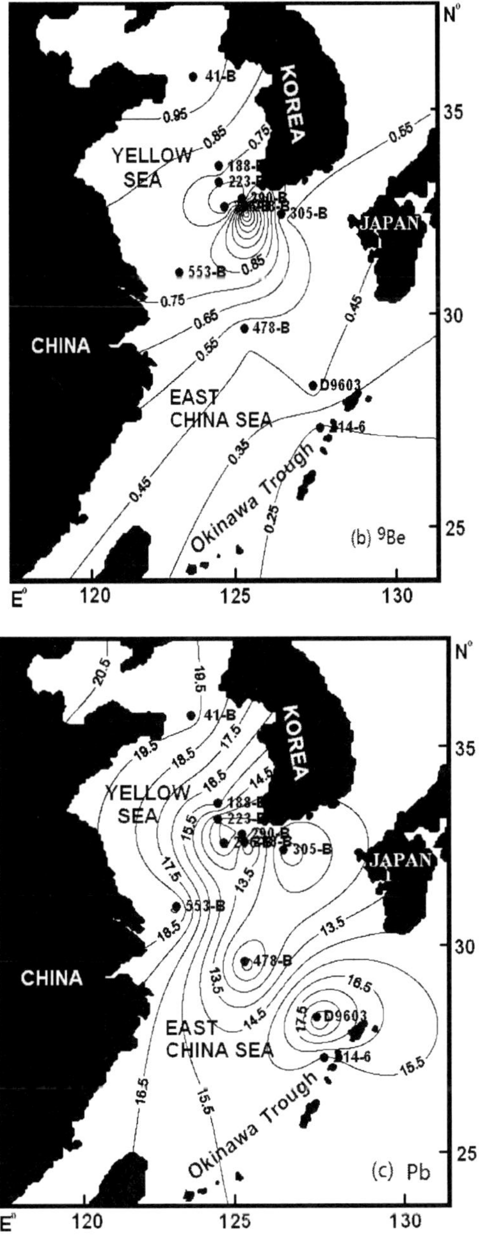

Figure 8. Distribution patterns of (a) ^{10}Be, (b) ^{9}Be, and (c) Pb in the surface sediments from the Yellow Sea and the northern East China Sea continental shelves and the northern Okinawa Trough (units for the figures in the contour lines are $\times 10^8$ atoms g^{-1} (^{10}Be) and g g^{-1} (^{9}Be and Pb), respectively.

CONCLUSION

(1) The seawater profiles of ^{10}Be presented here have given us a first look at the ^{10}Be distribution in a continental shelf influenced by one of the world largest rivers and the Kuroshio Current. Dissolved ^{10}Be concentrations in sea water of the East China Sea and the Okinawa Trough are mainly controlled by the surface biological productivity, particle remineralization, and the degree of mixing with the Yangtze River and the Kuroshio waters. During the sampling periods (summer and autumn), the East China Sea was well stratified. Generally, the ^{10}Be water depth profiles can be divided into three layers: the surface mixed layer, the particulate ^{10}Be regeneration layer, and the bottom layer. The ^{10}Be concentrations in surface water increase gradually towards the Kuroshio and increase sharply at the edge of the Kuroshio Current. Vertical distributions of ^{10}Be in water columns showed that in the summer ^{10}Be was enriched in the bottom water near the Yangtze River estuary and the bottom water in the middle of the continental shelf. The two enriched areas were separated, probably by an intrusion of the Continental Coastal Water. In the autumn, ^{10}Be bottom enrichment only occurred in the western part of the East China Sea. This phenomenon is consistent with the seasonal circulation pattern change of currents induced by monsoon winds. ^{10}Be is introduced to the East China Sea through atmospheric precipitation, river discharge, and the Kuroshio Current, and the input flux of ^{10}Be is much higher than that in the open ocean. Simple box model results indicate that ^{10}Be input from the Kuroshio Current is more important than the Yangtze River input and atmospheric precipitation. About 81% of ^{10}Be input to the East China Sea is scavenged into the sediments and 19% of ^{10}Be flows out of the East China Sea by currents and water exchange. The ^{10}Be sedimentation flux in the East China Sea is nearly five times of the average global ^{10}Be production rate, suggesting that the East China Sea is an important sink for ^{10}Be.

(2) Variations in ^{10}Be concentrations in sediment cores from the Okinawa Trough can be attributed either to changes in paleoclimate and paleoceanography (sea levels, biological productivity in euphotic zones, current circulations etc.) and to variations in the production rate of ^{10}Be due to changes in the Earth's magnetic field. Through comparisons with the ^{10}Be records from ice cores in which ^{10}Be concentrations reflect atmospheric ^{10}Be production rates more

directly, records of ^{10}Be in marine sediment can be used to depict the influences from climate changes on ocean currents. The ^{10}Be records in the Okinawa Trough sediment during the last glacial period, the Younger Dryas cold period, and the Holocene showed that the main part of the fluctuations in ^{10}Be concentrations in the sediment can be attributed to variations in the ^{10}Be production rate. The high ^{10}Be concentrations and flux in the Okinawa Trough during the glacial period proved that on millennium scale the Kuroshio Current still flowed in the Okinawa Trough area with significantly high intensity. However, this does not disprove that short changes in intensity or direction of the Kuroshio ever occurred during certain short cooling events. There were clear responses of ^{10}Be concentration and flux in core 03 on short-term climate events which may have caused the changes in the strength and the flow path of the Kuroshio. The ^{10}Be minimum during the Younger Dryas event may indicate that the Kuroshio had been once greatly weakened or even terminated in the Okinawa Trough area which could be related to the response of the Pacific Ocean to the Younger Dryas.

(3) The ^{10}Be distribution in surface sediments in the study area showed a decreasing trend in concentration level: the northern Okinawa Trough> the South Yellow Sea mud area> the East China Sea, indicating that the ^{10}Be distribution is mainly affected by the flow of the Kuroshio and its branches, and the input of ^{10}Be from the HWC in the Yellow Sea can not be ignored. The ^{9}Be distribution in surface sediments showed a decreasing trend: the South Yellow Sea mud area> the East China Sea continental shelf>the northern Okinawa Trough. The Pb distribution in surface sediments are characterized by a decreasing concentration trend: the northern Okinawa Trough> the East China Sea continental shelf> the South Yellow Sea mud area. Entirely different distribution configurations of ^{10}Be, ^{9}Be, and Pb generally reflect the different importance of influences of the river input and the Kuroshio on ^{10}Be, ^{9}Be, and Pb in sediments.

The ^{10}Be budget in seawater in the East China Sea shows that ^{10}Be input from the Kuroshio contribution to the East China Sea is almost an order of magnitude higher than the contribution of the Yangtze River. The Kuroshio plays a role of a conveyor belt which carrying ^{10}Be from the ocean into the East China Sea and the Okinawa Trough area. ^{10}Be can therefore be used as a tracer for the Kuroshio. The ^{10}Be records in sediment cores from the Okinawa Trough should reflect the Kuroshio evolution history since the last glacial period. From the discussions above, it can be seen that ^{10}Be can effectively used as a geochemical tracer in paleoceanographical and paleoclimatological studies for marginal seas.

ACKNOWLEDGMENT

This work was supported by the National Science Foundation of China (Grant Nos. 41073011, 40173027, 40176018, 90411014, and 4042115001). We thank H. Matsuzaki of Micro Analysis Laboratory, Tandem Accelerator (MALT), the University of Tokyo, Japan, for AMS analysis.

REFERENCES

Alley, RB; Meese, DA; Shuman, CA; Gow, AJ; Taylor, KC; Grootes, PM; White, JWC; Ram, M; Waddington, ED; Mayewski, PA; Zielinski, GA. Abrupt increase in Greenland snow accumulation at the end of the Younger Dryas event. *Nature*, 1993, 362, 527–529.

Anderson, RF; Lao, Y; Broecker; WS; Trumbore, SE; Hofmann, HJ; Wolfli, W. Boundary scavenging in the Pacific Ocean: A comparison of ^{10}Be and ^{231}Pa. *Earth and Planetary Science Letters*, 1990, 96, 287–304.

Alexander, CR; DeMaster, DJ; Nitrouer, CA. Sediment accumulation in a modern continental shelf setting: The Yellow Sea. *Marine Geology*, 1991, 98, 51–72.

Beardsley, RC; Limburner, R; Yu, H; Cannon, GA. Discharge of the Changjiang (Yangtze River) into the East China Sea. *Continental Shelf Research*, 1985, 4, 57–76.

Broecker, WS; Andrge, M; Wolfli, W; Oeschger, H; Bonani, G; Kennett, J; Peteet, B. The chronology of the last deglaciation: Implication to the cause of the Younger Dryas Event. *Paleoceanography*, 1988, 3(1), 1–19.

Cai, DL; Shi, XF; Zhou, WJ; et al. Sources and transportation of the materials consisted of suspended matter and sediment in the southern Yellow Sea. *Chinese Science Bulletin*, 2001, 46 (Suppl.), 16–23.

Chen, CTA; Ruo, R; Pai, SC; Liu, CT; Wong, GTF. Exchange of water masses between the East China Sea and the Kuroshio off northeastern Taiwan. *Continental Shelf Research*, 1995, 15, 19–39.

Chen, CTA. The Kuroshio intermediate water is the major source of nutrients on the East China Sea continental shelf. *Oceanologica Acta*, 1996, 19, 523–527.

Cheng, ZP; Shi, XF; Wu, Y, H; Su, X; Li, XY; Wang, KS; Yang, YL; Ge, SL; Ju, XH; Shi, FD; Matsuzaki, H. Growth ages of ferromanganese crusts from the western and central Pacific: Comparison between nannofossil analysis and ^{10}Be dating. *Chinese Science Bulletin*, 2006, 51(24), 3035–3040.

Christl, M; Strobl, C; Mangini, A. Beryllium-10 in deep-sea sediments: A tracer for the Earth's magnetic field intensity during the last 200000 years. *Quaternary Science Reviews*, 2003, 22, 725–739.

Elenga, H; Peyron, O; Bonnefille, R; Jolly, D; Cheddadi, R; Guiot, J; Andrieu, V; Bottema, S; Buchet, G; de, Beaulieu, J-L; Hamilton, A, C; Maley, J; Marchant, R; Perez-Obiol, R; Reille, M; Riollet, G; Scott, L; Straka, H; Taylor, D; Van, Campo, E; Vincens, A; Laarif, F; Jonson, H. Pollen-based biome reconstruction for southern Europe and Africa 18,000 yr BP. *Journal of Biogeography*, 2000, 27, 621–634.

Frank, M; Eisenhauer, A; Bonn, WJ; Walter, P; Grobe, H; Kubik, PW; Dittrich-Hannen, B; Mangini, A. Sediment redistribution versus paleoproductivity change: Weddell Sea margin sediment stratigraphy and biogenic particle flux of the last 250,000 years deduced from $^{230}Th_{ex}$, ^{10}Be and biogenic barium profiles. *Earth and Planetary Science Letters*, 1995, 136, 559–573.

Hu, FS; Kaufman, D; Yoneji, S; Nelson, D; Shemesh, A; Huang, YS; Tian, J; Bond, G; Clegg, B; Brown, T. Cyclic variation and solar forcing of Holocene climate in the Alaskan Subarctic. *Science*, 2003, 301(5641), 1890–1893.

Finkel, R, C; Nishiizumi, K. Beryllium-10 concentrations in the Greenland Ice Sheet Project 2 ice core from 3~40 ka. *Journal of Geophysical Research*, 1997, 102 (C12), 26669–26706.

Gong, GC; Chen, YLL; Liu, KK. Chemical hydrography and chlorophyll a distribution in the East China Sea in summer: implications in nutrient dynamics. *Continental Shelf Research*, 1996, 16, 1561–1590.

Guo, ZG; Yang, ZS; Chen, ZL, Mao, D. Sources of sedimentary organic matter in the mud areas of the East China Sea shelf. *Geochimica*, 2001, 30(5), 416–424. (In Chinese with English abstract)

Hartz, I; Milthers, VD. Det senglaciale Ler i Allerød Teglværksgrav. *Meddelelser fra Dansk geologisk Foreningen*, 1901, 8, 31–60.

Heikkila, U; Beer, J; Feichter, J. Meridional transport and deposition of atmospheric ^{10}Be. *Atmos. Chem. Phys.*, 2009, 9, 515–527.

Jian, Z; Wang, P=X; Saito, Y; Wang, JL; Pflaumann, U; Oba, T; Cheng, XR. Holocene variability of the Kuroshio Current in the Okinawa Trough, northwestern Pacific Ocean. *Earth and Planetary Science Letter*, 2000, 184, 305–319.

Johnsen, SJ; Clausen, HB; Dansgaard, W; Fuhrer, K; Gundestrup, N; Hammer, CU; Iversen, P; Jouzel, J; Stauffer, B; Steffensen, JP. Irregular glacial interstadials recorded in a new Greenland ice core. *Nature*, 1992, 359, 311–313.

Kim, YL. *Marine Geology of the East China Sea*. Ocean Press, Beijing, 1992, 524 pp. (In Chinese with English abstract).

Ku, TL; Kusakabe, M; Measures, CI; Southon, JR; Cusimano, G; Vogel, JS; Nelson, DE; , Nakaya, S. Beryllium isotope distribution in the western North Atlantic: A comparison to the Pacific. *Deep-Sea Research*, 1990, 37, 795–808.

Kusakabe, M; Ku, TL; Southon, JR; Liu, S; Vogel, JS; Nelson, DE; Nakaya, S; Cusimano, G. Be isotopes in rivers/estuaries and their oceanic budget. *Earth and Planetary Science Letters*, 1991, 102, 265–276.

Kusakabe; M; Ku; TL; Vogel; JS; Southon; JR; Nelson; DE; Richards; G. ^{10}Be profiles in seawater. *Nature*, 1982, 299, 712–714.

Kusakabe, M; Ku, TL; Southon, JR; Vogel, JS; Nelson, DE; Measures, CI; Nozaki, Y. Distribution of ^{10}Be and ^{9}Be in the Pacific Ocean. *Earth and Planetary Science Letters*, 1987, 82, 231–240.

Lan, DZ; Chen, CH; Li, C. Sedimentary diatom records on excursion of Okinawa current in Okinawa Trough since last glaciation. *Acta Palaeontologica Sinica*, 2003, 42(3), 466–472.

Laj, CM; Mazaud, A; Duplessy, JC. Geomegnetic intensity and ^{14}C abundance in the atmosphere and ocean during the past 50 kyr. *Geophysical Research Letters*, 1996, 23, 204–2048.

Lao, Y; Anderson, RF; Broecker, WS; Trumbore, SE; Hofmann, HJ; Wolfli, W. Transport and burial rates of ^{10}Be and ^{231}Pa in the Pacific Ocean during the Holocene period. *Earth and Planetary Science Letters*, 1992, 113(12), 173–189.

Lao, Y; Anderson, RF; Broecker, WS; Trumbore, SE; Hofmann, HJ; Wolfli, W. Increased production of cosmogenic ^{10}Be during the Last Glacial Maximum. *Nature*, 1992, 357, 57–578.

Li, TG; Liu, ZX; Hall, MA; Berne, S; Saito, Y; Cang, SX; Chen, ZB. Heinrich event imprints in the Okinawa Trough-evidence from oxygen isotope and planktonic foraminifera. *Palaeogeography, Palaeoclinatology, Palaeoecology*, 2001, 176, 133–146. (In Chinese with English abstract)

Li, TG; Jiang, B; Sun, RT; Zhang, DY; Liu, ZX; Li, Q. Evolution pattern of warm current system of the East China Sea and the Yellow Sea since the last deglatiiation. *Quat. Sci.*, 2007, 27(6), 945–954. (In Chinese with English abstract)

Lin, CY; Shyu, CZ; Shih, WH. The Kuroshio fronts and cold eddies off northeastern Taiwan observed by NOAA-AVHRR imageries. Terrestrial, *Atmospheric and Oceanic Sciences*, 1992, 3, 225–242.

Liu, ZX; Li, PY; Li, TG; Huang, QY; Cheng, ZB; Wei, GL; Liu, LJ; Li, Z; Berne, S; Saito, Y. Paleoclimate events in Okinawa Trough since 50,000 a BP *Chinese Science Bulletin*, 2000, 45(16), 883–887.

Meng, XW; Du, DW; Liu, YG; Liu, ZX. Molecular bio-marker record of paleooceanographic environment in the East China Sea during the last 35000 years. *Science in China* (*Series D*), 2002, 45, 184–192.

Miao; Y; Su; JL; Yu; H. Mixing feature of water type in the East China Sea in summer. In: Su, J.L. (Ed.), Selected Papers on Kuroshio Research. Ocean Press, Beijing, 1987, 204–217 (In Chinese with English abstract).

McHargue, LR; Damon, PE. The global beryllium-10 cycle. *Rev. Geophys.*, 1991, 29, 141–158.

Monaghan, MC; Krishnaswami, S; Turekian, KK. The global-average production rate of ^{10}Be. *Earth and Planetary Science Letters*, 1985, 76, (3-4), 279–287.

Muscheler, R; Beer, J; Wagner, G; Finkel, RC. Changes in deep-water formation during the Younger Dryas event inferred from ^{10}Be and ^{14}C records. *Nature*, 2000, 408, 567~570.

Nozaki, Y; Kasemsupaya, V; Tsubota, H. Mean residence time of the shelf water in the East China and the Yellow Seas determined by 228Ra/226Ra measurements. *Geophysical Research Letters*, 1989, 16, 1297–1300.

Olsen, CR; Larsen, IL; Lowry, PD; Cutshall, NH. Geochemistry and deposition of ^{7}Be in river-estuarine and coastal waters. *Journal of Geophysical Research*, 1986, 91, 896–908.

Park, YA; Khim, BK. Origin and dispersal of recent clay minerals in the Yellow Sea. *Marine Geology*, 1992, 104, 205–213.

Raisbeck, GM; Yiou, F; Fruneau, F; Loiseaux, JM; Lieuvin, M. ^{10}Be concentration and residence time in the ocean surface layer. *Earth and Planetary Science Letters*, 1979, 43, 237–240.

Raisbeck, GM; Yiou, F; Fruneau, F; Loiseaux, JM; Lieuvin, M; Ravel, JC; Reyss, JM; Guichard, F. ^{10}Be concentration and residence time in the deep ocean. *Earth and Planetary Science Letters*, 1980, 51, 275–278.

Saito, Y; Yang, Z-S. The sediment budget in the East China Sea. Proceedings of the Third Symposium on Geoenvironments and Geo-technics, Committee of Environmental Geology of the Geological Society of Japan, 1993.

Shen, CD; Beer, JP; Kubik, W; Suter, M; Borkovec, M; Liu, TS. Grain size distribution, ^{10}Be concentrations and magnetic susceptibility of micrometer-nanometer loess materials. *Nuclear Instruments and Methods in Physics Research Section B*, 2004, 223, 613–617.

Shen, SX; Chen, LR; Gao, L; Li, AC. Discovery of Holocene cyclonic eddy sediment and pathway sediment in the southern Yellow Sea. *Oceanologia et Limnologia Sinica*, 1993, 24(6), 563–570. (In Chinese with English abstract)

Shieh, YT; Chen, MP. The ancient Kuroshio Current in the Okinawa Trough during the Holocene, *Acta Oceanogr.Taiwan*, 1995, 34 (4), 73–80.

Somayajulu, BLK; Sharma, P; Beer, J; Bonani, G; Hofmann, HJ; Morenzonri, E; Nessi, M; Suter, M; Woelfli, W. ^{10}Be annual fallout in rains in India. *Nuclear Instruments and Methods in Physical Research*, 1984, B5, 398–403.

Sun, X. Analysis of the surface path of the Kuroshio in the East China Sea. In: Sun, X. (Ed.), *Essays on the Investigation of Kuroshio*. Ocean Press, Beijing, 1987, 1–14 (In Chinese with English abstract).

Thompson, PR. Planktonic foraminifera in the western North Pacific during the past 150,000 year: comparison of modern fossil assemblages. *Palaeogeography, Palaeoclinatology, Palaeoecology*, 1981, 35, 241–279.

Ujiié, H; Ujiié, Y. Late Quaternary course changes of the Kuroshio Current in the Ryukyu Arc region, northwestern Pacific Ocean. *Marine Micropaleontology*, 1999, 37, 23–40.

von Blanckenburg, F; O'Nions, RK; Belshaw, NS; Gibb, A; Hein, JR. Global distribution of beryllium isotopes in deep ocean water as derived from Fe-Mn crusts. *Earth and Planetary Science Letters*, 1996, 141(1~4), 213–226.

Wang, PX. The ice-age China Sea-research results and problems. In: Wang Pinxian, Lao Qiuyuan, He Qixiang eds. *Proceedings of the First International Conference on Asian Marine Geology*. Beijing, China Ocean Press, 1990, 181–197.

Wang, PX. Response of Western Pacific marginal seas to glacial cycles: Paleoceanographyic and sedimentological features. *Marine Geology*, 1999, 156, 5–39.

Wang, PX. Response of Western Pacific marginal seas to glacial cycles: paleoceanographic and sedimentological features. *Marine Geology*, 1999, 156, 5–39.

Wang, YJ; Wang, YJ; Cheng, H; Edwards, RL; An, ZS; Wu, JY; Shen, CC; Dorale, JA. A high-resolution absolute-dated late Pleistocene monsoon record from Hulu Cave, China. *Science*, 2001, 294, 2345–2348.

Xiong, YQ; Liu, ZX. Variations in sediment provenance and its implications of Core DGKS9603 since the Late Quaternary. *Acta Oceanologica Sinica*, 2004, 26, (2), 61-71. (In Chinese with English abstract)

Xu, XD; Oda, M. Surface-water evolution of the eastern East China Sea during the last 36,000 years. *Marine Geology*, 1999, 156, 285–304.

Yan, J; Cang, SX. Study on oxygen isotope Stratigraphy in core Z14-6 from Okinawa Trough. *Oceanologia et Liminologia Sinica*, 1990, 21(5), 442–448. (In Chinese with English abstract)

Yanagi, T; Takahashi, S; Hoshika, A; Tanimoto, T. Seasonal variation in the transport of suspended matter in the East China Sea. *Journal of Oceanography*, 1996, 52, 539–552.

Yanagi; T. Material transport in the Yellow/East China Seas. *Bulletin on Coastal Oceanography*, 1994, 31, 239–256. (In Japanese with English abstract)

Yang, YL; Kusakabe, M; Southon, JR. ^{10}Be profiles in the East China Sea and the Okinawa Trough. *Deep-Sea Research Part II: Topical Studies in Oceanography*, 2003, 50(2), 339–351.

Yang, ZS; Guo, ZG; Wang, ZX; Xu, JP; Gao, WS. Macro-pattern of suspended sediment transportation in the Yellow Sea and the East China Sea towards the eastern deep-sea areas. *Acta Oceanologica Sinica*, 1992, 14(2), 81–90. (In Chinese with English abstract)

Yi, WX; Shen, CD; Zhong, HH; Hu, GH; Liu, TS. High resolution element records in the late Pleistocene Xifeng loess profile. *Chinese Journal of Geochemistry*, 1996, 15(3), 272–277. (In Chinese with English abstract)

Zhou, HY; Li, TG; Jia, GD; Zhu, ZY; Chi, BQ; Cao, QY; Sun, RT. Sea surface temperature reconstruction with long-Chian unsaturated alkenones for the middle Okinawa Trough during the last glacial- interglacial cycle. *Oceanologia et Liminologia Sinica*, 2007, 38(5), 1–8. (In Chinese with English abstract).

In: Beryllium
Editor: Pauleen Dyer

ISBN: 978-1-63321-590-0

Chapter 4

BERYLLIUM-7 CONTENT IN RAIN: EVIDENCES FOR A SEMIARID ENVIRONMENT

***J. Juri Ayub*[1*]*, R. H. Velasco*[1]*, M. Rizzotto*[1]*
and R. M. Anjos*[2]

[1]Grupo de Estudios Ambientales, Instituto de Matemática Aplicada San Luis, Universidad Nacional de San Luis/ CCT-San Luis, CONICET, San Luis Argentina

[2] Laboratório de Radioecologia e Alterações Ambientais, Instituto de Física, Universidade Federal Fluminense, Niterói, RJ, Brazil

ABSTRACT

Beryllium-7 (^{7}Be) is a relatively short-lived radionuclide (half-life 53.3 days) which decays by electron capture either directly to the ground state of ^{7}Li (89.56%) or to an excited state of ^{7}Li (10.44%), which decays to the ground state of ^{7}Li via gamma-ray emission at 477.6 keV. This allows us to easily quantify it by using gamma-ray spectrometers. Beryllium-7 has a cosmogenic origin and is produced in the upper atmosphere and lower stratosphere by high-energy spallation interactions of nitrogen and oxygen. It continuously enters to marine and terrestrial ecosystems via wet (over 90%) and dry (3 to 10%) deposition. Several factors can affect this input, such as production rate (which varies with latitude, altitude, and solar activity), stratosphere–troposphere mixing,

* E-mail: jjuri@unsl.edu.ar.

circulation and advection processes within the troposphere and efficiency with which it is removed from the troposphere. After deposition, ^{7}Be will tend to associate with particulate material (particle-reactive element). Its relatively short half-life, reactivity, and continuous and definable production rates make ^{7}Be a potentially powerful tool for the study and description of several environmental processes such as soil redistribution, sediment sources assessment, concentration in air, air mass transport, study of metal scavenging and others. In order to use ^{7}Be as an environmental tracer, the knowledge of its input from the atmosphere and its variability are needed. However, when its input is evaluated, divergent information may be obtained. For different regions, dissimilar environmental conditions and seasons of the year, ^{7}Be rain water content shows a high variability, and the cause of these changes could be difficult to understand or explain. A high ^{7}Be content has been reported for some environments for precipitations of a few millimeters and low ^{7}Be contents for precipitations occurring after other precipitation event. These results have been explained by the atmospheric washing phenomenon and a reload rate can be estimated. Moreover, effects of rainfall rate on rain ^{7}Be content have been reported with divergent results. Despite these, there is agreement that wet deposition on the ground can be estimated from the rainfall volume. This chapter summarizes the results obtained in evaluating the ^{7}Be content in rainfalls for a semiarid environment characterized by a seasonal precipitation regime. For entire rain events, the effect of precipitation variables on ^{7}Be content in rain water is evaluated and contrasted with other regions. For single rain events the changes of ^{7}Be content and the effect of rainfall intensity is evaluated for each millimeter of rain fallen.

INTRODUCTION

Beryllium is present in the environment and is formed by the stable isotope Beryllium-9 and two cosmogenic isotopes, Beryllium-10 and Beryllium-7 (^{7}Be), which are of interest in earth studies. The focus of this chapter is the ^{7}Be content in rainfalls.

Beryllium-7 is a relatively short-lived radionuclide, with a half-life of 53.3 days, while ^{10}Be has a half-life of 1.5×10^{6} years. Beryllium-7 decays by electron capture either directly to the ground state of ^{7}Li (89.56%) or to the excited state of ^{7}Li (10.44%), which decays to the ground state of ^{7}Li via gamma-ray emission at 477.6 keV. This radionuclide is produced mainly in the stratosphere (~70%) and the remaining part in the troposphere (Johnson & Viezee, 1981; Cannizzaro et al., 2004) by high-energy spallation of cosmic rays with nucleus of nitrogen and oxygen which are present in the terrestrial atmosphere (Lal et al., 1958). The nuclear reaction produces BeO or $Be(OH)_2$ which diffuse through the atmosphere and adsorb electrostatically to atmospheric aerosols, thus entering the ecosystems.

The production of this radionuclide depends on the cosmic rays flux, which varies with latitude, altitude and solar activity. It is recognized that the production in the atmosphere increments from the equator to the poles (Kaste et al., 2002) and is higher between 12 and 20 km altitude, decreasing exponentially to the ground surface (Lal el al, 1958; Bhandari et al, 1970). Moreover, the cosmogenic Beryllium production varies with the 11-year solar cycle. Solar activity maximums result in increased deflection of cosmic rays from the solar system that decreases the cosmic-ray flux to the earth, and thus decreases ^{7}Be production (Kaste et al., 2002; Cannizzaro et al., 2004). Therefore, factors that influence the concentration in the atmosphere include stratosphere–troposphere mixing, circulation and advection processes within the troposphere and the efficiency with which it is removed from the troposphere (Feely et al., 1989; Kaste et al., 2002).

Beryllium-7 is removed from the atmosphere by radioactive decay and wet and dry deposition. Dry deposition represents less than ten percent of the total fallout, being the remainder deposition accounted for by rainout (raindrop formation around ^{7}Be carrier particles) and washout (^{7}Be carriers gathered by falling rain), see Murray et al. (1992), Ishikawa et al. (1995), Wallbrink & Murray (1994), Salisbury & Cartwright (2005), Ioannidou et al. (2005) and Kaste et al. (2002). By subsequent deposition ^{7}Be reaches the soil surface and is rapidly and strongly fixed to the fine soil particles (Kaste et al., 2002; Andrello & Appoloni, 2010).

Beryllium-7 has shown to be a useful tracer in several atmospheric, geochemical, erosional and sedimentological processes, such as atmospheric transport, air mass sources, vertical transport within the troposphere, metal scavenging, resuspended sedimentary material in coastal water and more recently, quantification of sedimentation and erosion.

BERYLLIUM-7 IN RAINS

As mentioned before, ^{7}Be enters the ecosystems mainly by wet deposition, therefore measurements of ^{7}Be content in precipitations provide a good way to quantify the total amount that comes from the atmosphere.

With the intention of using this radionuclide as an environmental tracer, several studies tending to understand the factors that could affect the ^{7}Be income by rainfalls and the total content in the soil profile have been performed in different regions and for dissimilar environments.

These studies have been performed using different methodologies for rain water collection and different treatments on this collected water were carried out with the aim of obtaining the best quantification of ^{7}Be content in rainfalls. For the rain water collection specific devices were built or acquired according to whether the study involves the collection of all rain falling over a period of time (weekly or monthly) or the collection of single rain events. After collection some techniques of purification and concentration of ^{7}Be may be necessary, such as the use of cations exchange columns, coprecipitation of Be with Fe hydroxides and other hydroxide phases. On the other hand, the high-purity germanium detectors currently available enable the determination of ^{7}Be without sample treatment. A more detailed information about sample collection and treatment are described in Kaste et al. (2002), Caillet et al. (2001) and Juri Ayub et al. (2009 and 2012). Despite the different methodologies for collection and treatment of rain water samples most studies give comparable results.

Table 1. The ^{7}Be activity concentration in rainfalls at different locations

Location	precipitation (mm)		^{7}Be (Bq l^{-1})		Period	Reference
	min	max	min	max		
Argentina (San Luis)	1	111	0.4	3.2	11/2006 - 03/2009	Juri Ayub et al., 2012
Australia (Canberra)	0.1	90	0.02	5.9	04/1988 - 08/1989	Wallbrink & Murray, 1994
Czech Republic	*	*	0.6	1.9	05/2005 - 10/2007	Pöschl et al., 2010
Monaco	*	*	0.04	8.6	1998 - 2010	Pham et al., 2013
Spain (Huelva)	1.2	227	0.03	7.42	04/2009 – 07/2010	Lozano et al., 2011
Spain (Málaga)	5.6	221.7	0.55	6.1	01/1992-12/1999	Dueñas et al., 2001
Switzerland	0.2	66	0.93	10.45	11/1997 - 11/1998	Caillet et al., 2001
United Kingdom (Chilton)	*	*	0.59	2.74	08/1959 - 08/1961	Peirson, 1963
USA	*	*	0.5	1.59	*	Brown et al., 1988
USA (California)	7	140	1.3	4.4	10/2009 - 08/2010	Conaway et al., 2013
USA (College Station - Texas)	*	*	0.26	5.0	06/1989 - 05/1992	Baskaran et al., 1993
USA (Galveston - Texas)	*	*	0.09	20.7	12/1988 - 02/1992	Baskaran et al., 1993
USA (Massachusetts)	*	*	0.35	24.03	03/1996 - 02/1998	Benitez-Nelson & Buesseler, 1999
USA (Michigan)	5.6	95	1.26	10.9	09/1999 - 02/2001	McNeary & Baskaran, 2003
USA (Virginia)	*	*	0.48	3.18	*	Todd et al., 1989

* no available information.

The reported ^{7}Be content in rainfalls in the literature show a wide range of values which make clear its variability with the geographical location of the study sites (Table 1). On the other hand, for the same site, changes between two and three orders of magnitude in rainwater ^{7}Be content are recorded.

These variations could come from the time taken by each study. Several authors have reported seasonal variations in ^{7}Be content in rainfall, with higher values in spring and summer than in autumn and winter, derived from the intrusion into the troposphere of stratospheric air masses rich in ^{7}Be, in the warmer months of the year. The study performed by Pham et al. (2013) covers one complete 11-year solar cycle, but the oscillation on ^{7}Be content in rainfalls could not be attributed to a dependence with the solar cycle, as was clearly shown for the atmospheric ^{7}Be content by Pham et al. (2011) and Cannizzaro et al. (2004).

There is a general agreement that changes in ^{7}Be content as a function of time in rainfalls could be expected because of fluctuations of its atmospheric content. Nevertheless, for a semi-arid environment, the measurements of ^{7}Be content in rainfalls performed from November 2006 to March 2009 showed no significant differences between the months of the year (Figure 1), despite the seasonal rainfalls distribution.

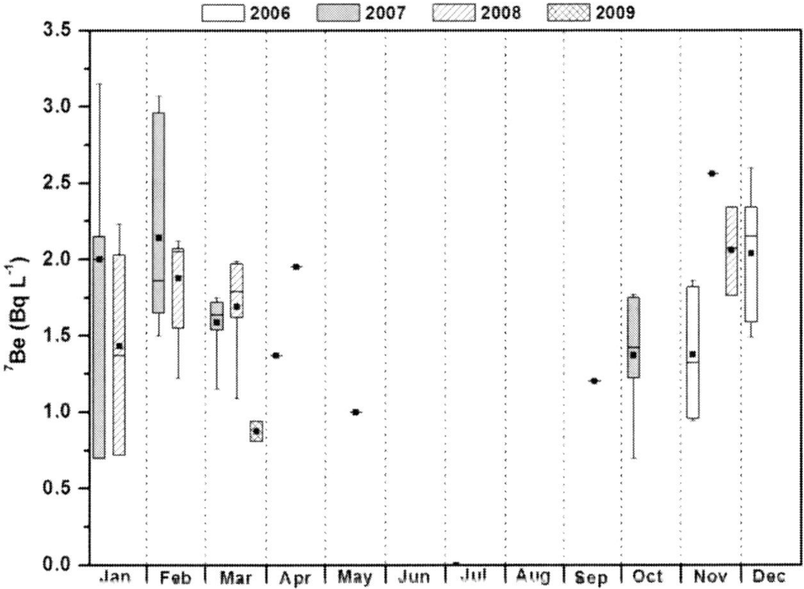

Figure 1. Box chart representation of ^{7}Be activity concentration in rainwater for rainfall events clustered according to the month of the year. Boxes are determined by the 25th and 75th percentiles. Whiskers are determined by the 5th and 95th percentiles. The arithmetic means of ^{7}Be activity concentration values are also represented. Modified from Juri Ayub et al., 2012.

Several studies have been performed with the aim of understanding the variations of ^{7}Be content in rains and the factors affecting them. The most important factors that arise are rainfall magnitude, rainfall intensity, elapsed time between rain events and rain duration.

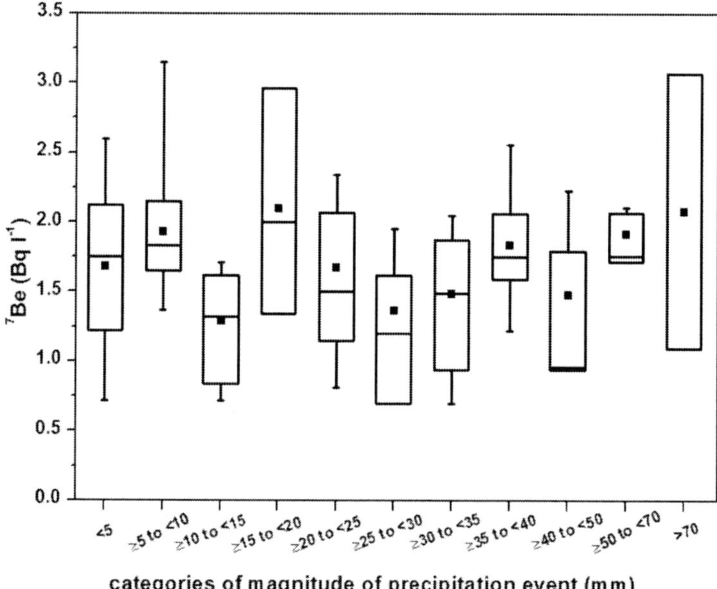

Figure 2. Box chart representation of ^{7}Be activity concentration in rainwater for rainfall events clustered according to precipitation magnitude. Boxes are determined by the 25th and 75th percentiles. Whiskers are determined by the 5th and 95th percentiles. The arithmetic means of ^{7}Be activity concentrations values are also represented. Extracted from Juri Ayub et al., 2012.

When ^{7}Be content in rain is analyzed based on the precipitations magnitude, two different behaviors are found. On one hand, precipitations of a few millimeters show higher values in ^{7}Be content than precipitations of higher magnitude (Baskaran et al., 1993; McNeary & Baskaran, 2003; Caillet et al., 2001; Pham et al., 2013). Caillet et al. (2001) found two different rainfall groups based on its ^{7}Be content; the highest values are found in low rainfalls with a break in 20 mm of rain and Baskaran et al. (1993) and McNeary & Baskaran (2003) established a negative dependence between ^{7}Be content and magnitude of precipitation. On the other hand, no relationship between ^{7}Be content and precipitation is reported. Benitez-Nelson & Buesseler (1999) reported some rains of a few millimeters (<5mm) with high ^{7}Be content, however they did not find a negative relation between these two variables. Conaway et al. (2013) and Juri Ayub et al (2012) also found no changes in ^{7}Be which could be attributed to rain magnitude. On the other hand, Wallbrink & Murray (1994) did not find a significant correlation between ^{7}Be content and rainfall magnitude for precipitation events less than 25 mm. Figure 2 shows the ^{7}Be content registered by Juri Ayub et al (2012) for individual events clustered according to the categories of rain magnitude. As can be observed there are no significant differences in ^{7}Be attributable to the volume of the event.

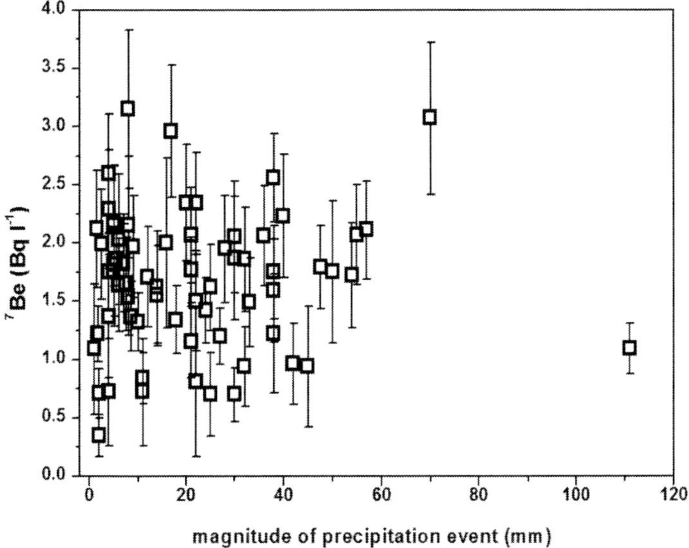

Figure 3. ^{7}Be activity concentration in rainwater for individual rain event versus precipitation magnitude. Error bars show the uncertainty in activity concentration measurements. Extracted from Juri Ayub et al., 2012.

These two different results could be conflicting. On one side the decreasing ^{7}Be content in rain with the increment of rainfall magnitude suggests that a washing phenomenon of atmosphere is taking place. The ^{7}Be production is assumed to be relatively constant within the troposphere at the same latitude, then the amount of ^{7}Be in the air column at that latitude may also remain approximately constant and the subtraction of this radionuclide by wet precipitation causes fewer ^{7}Be available through the precipitation event. This phenomenon should also dilute large-scale precipitations because the last millimeters of rain have lesser ^{7}Be content than the first millimeters. More information about the specific processes of ^{7}Be removal from the atmosphere may be found in Ischikawa et al. (1995). On the contrary, the studies where no changes in ^{7}Be content appear when rainfall amount increases, could suggest that in some regions the ^{7}Be content in the atmosphere is so great that no washing effect occurs. But if we analyze the reported individual data by Juri Ayub et al. (2012) (Figure 3) we find that one very great rainfall event (111 mm) recorded low ^{7}Be content in rainwater. For the study region, this event is the only one with this magnitude and is reported as an extreme and atypical rainfall. Nevertheless, this rainfall indicates that the washing effect could occur and that the threshold for this washing may be higher and dependent on the geographic location of sites.

Other three ways to evaluate if washing of the atmosphere occurs are:

analyze the ^{7}Be content in a rain which was preceded by another rain event, evaluate changes on ^{7}Be content through a single rain event and analyze the effect of duration of the rain on ^{7}Be content.

For the first case, Caillet et al. (2001), evaluating specifically ^{7}Be deposition from the atmosphere, report less deposition in rainfall preceded by another rain that took place 24 hours before. On the other hand, taking into account the elapsed time between two rain events (and season of the year) these authors estimate a mean time of 1.0 ± 0.1 day to reload the atmosphere with new ^{7}Be. In the second case, Ioannidou & Papastefanou (2006), analyzing sequential parts of a single event report changes in ^{7}Be content, which decreased by 2.5 or 3 times between the first and the last part of the rain.

For a semiarid environment in central Argentina, different results are reported (Juri Ayub et al., 2012). If the rainfall events are clustered in categories according to the elapsed time between events no changes on ^{7}Be content are recorded. In this case, a rainfall preceded by another rainfall that took place from a few hours (lesser than 1 day) to 15 days, showed similar ^{7}Be content (Figure 4). Taking into account that Caillet et al. (2001) report a reload rate of about one day, if the Argentinean data from rains that occurred in the time span of 24 hs or less are plotted, differences on ^{7}Be content are not found (Fig 5). The extreme rainfall of 111 mm was produced less than 24 hs after other rain and its ^{7}Be content is not smaller than the recorded for other rains, its value is in the expected range for the region. If washing effect from the atmosphere is relevant for the region, it should be observed in this rain because of its great magnitude and the short reload time. However, we must be cautious because is the only event with these features.

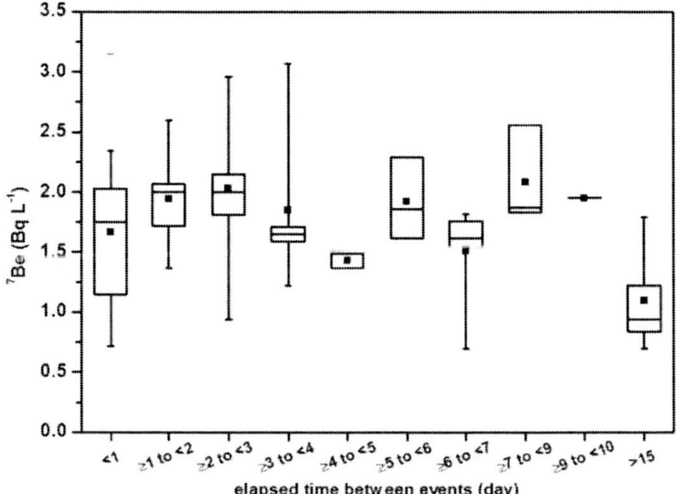

Figure 4. Box chart representation of ^{7}Be activity concentration in rainwater for rainfall events clustered according to the elapsed time between events. Boxes are determined by the 25th and 75th percentiles. Whiskers are determined by the 5th and 95th percentiles. The arithmetic means of ^{7}Be activity concentrations values are also represented. Extracted from Juri Ayub et al., 2012.

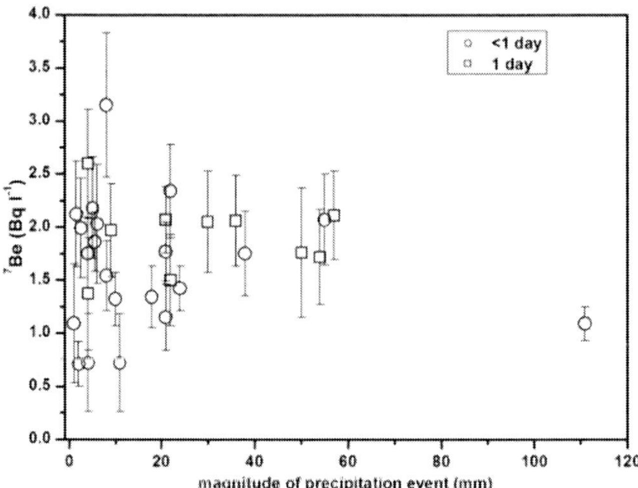

Figure 5. ^{7}Be activity concentration in rainwater for individual rain events with elapsed time of 24 hs (square) or less (circle) versus precipitation magnitude. Error bars show the uncertainty in activity concentration measurements.

Likewise, the lack of proof about washing effect in the Argentinean rainfall could derivate from a very short reload rate (a few hours) or a very high ^{7}Be atmospheric concentration, or both. Long-term studies are needed in order to know definitely the effect on ^{7}Be content in a rain due to a preceding one and the presence or absence of atmospheric washing.

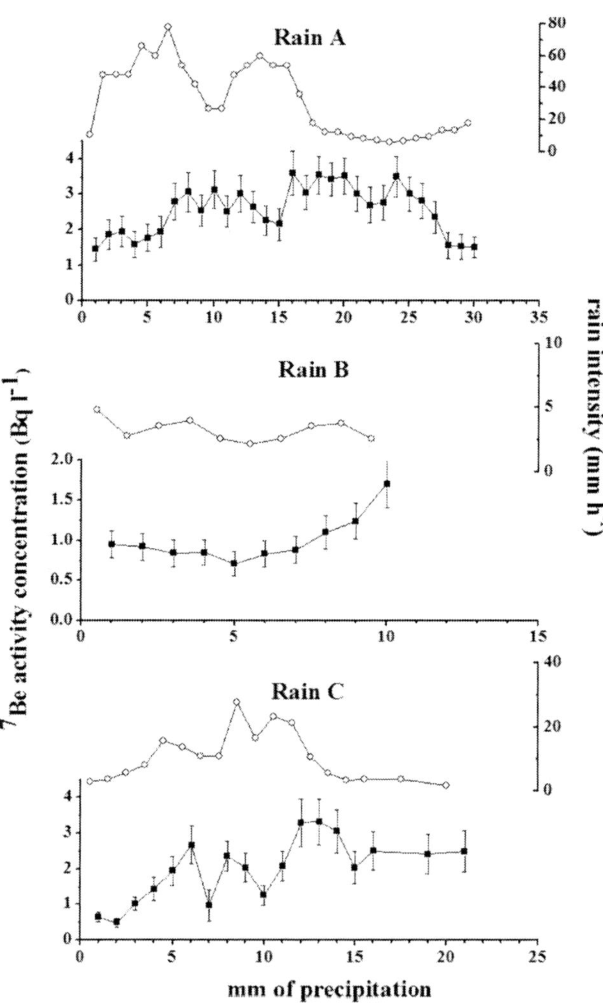

Figure 6. ^{7}Be activity concentration (filled square) and rain intensity (open circles) during a single precipitation event for (a) Rain A, (b) Rain B and (c) Rain C.

For the second case, three single rain events from Central Argentina were fractioned. Figure 6 shows the ^{7}Be activity concentration (filled squares) recorded for each millimeter of rain fallen. Rainfalls show variations in ^{7}Be content through the event; between the maximum and minimum values of ^{7}Be content we recorded changes of 2.5 times for the Rains A and B, while Rain C shows changes of 6.5 times across the event. The ^{7}Be content ranged from 1.45±0.33 Bq l^{-1} to 3.51±0.57 Bq l^{-1}, from 0.71±0.15 Bq l^{-1} to 1.71±0.30 Bq l^{-1} and from 0.49±0.12 Bq l^{-1} to 3.32±0.65 Bq l^{-1} for rains A, B and C, respectively. The mean values are in the range of the other rains: 2.6±0.4 Bq l^{-1}, 1.2±0.3 Bq l^{-1} and 2.1±0.4 Bq l^{-1} for Rains A, B and C, respectively. Despite the oscillations in ^{7}Be content through the rain, a clear pattern is not found. These rains clearly do not show high ^{7}Be contents in the first fallen millimeters.

While is expected that rains of few millimeters, like Rain B (10 mm) shows no a washing effect, we could expect that the other two rains (A: 30 mm, B: 21 mm) show a decrease in ^{7}Be content in the last millimeters if this radionuclide is washed in the atmosphere. This phenomenon does not appear in these data. Ioannidou & Papastefanou (2006) report that the first part of a rain removes the major amount of ^{7}Be and that the additional increase in the rainfall magnitude does not contribute to an additional ^{7}Be content in rainwater. Unfortunately these authors do not report the sequential ^{7}Be content across a single rain event, making difficult the comparison.

For the last case, Argentinean rainfalls can be clustered in several categories, according to the duration time of the rain (Figure 7). Very short rainfalls (lesser 1 hour) do not show different ^{7}Be contents than longer rainfalls, even rainfalls that have duration longer than 10 hours show similar contents. Inside this last category we also found a 24 hours duration rainfall that has very high ^{7}Be content ($\approx$ 3 Bq l^{-1}). This brings out the high variability that could be recorded in ^{7}Be content measures in rains and is a very important factor to be taken into account when we attempt to obtain global patterns about its behavior.

Another factor that has been suggested that may affect the ^{7}Be content in rain is the rainfall intensity. Figure 6 also shows the rain intensity (open circles) recorded for each millimeter of fallen rainfall. Although the correlation between rainfall intensity and ^{7}Be content is not found, the graphical picture seems to show that a high rainfall intensity is associated to a high ^{7}Be content. Ioannidou & Papastefanou (2006) report opposite results, since they find a high ^{7}Be content in rains when the rainfall intensity diminishes.

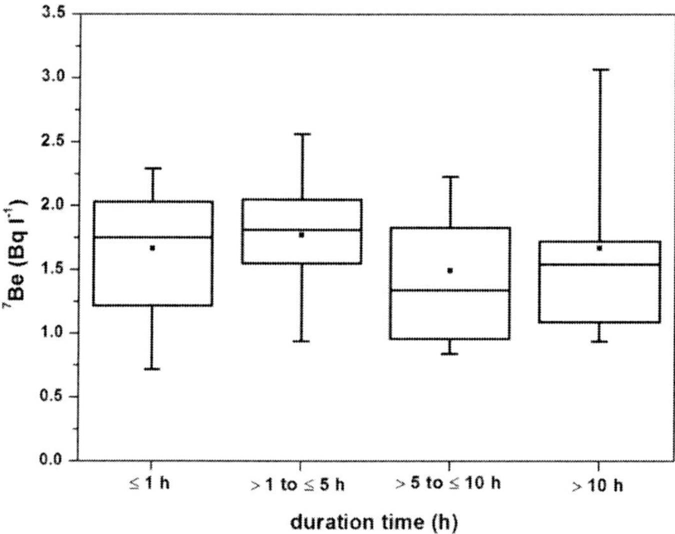

Figure 7. Box chart representation of ^{7}Be activity concentration in rainwater for rainfall events clustered according to duration time of the event. Boxes are determined by the 25th and 75th percentiles. Whiskers are determined by the 5th and 95th percentiles. The arithmetic means of ^{7}Be activity concentration values are also represented.

According to the mean rainfall intensity the rains can be arranged in:

light rain (< 2mm h^{-1}),
moderate rain (2 to 10 mm h^{-1}),
heavy rain (10 to 50 mm h^{-1}) and,
violent rain (> 50 mm h^{-1}).

In order to obtain a good statistics we clustered Argentinean rains in five categories of mean rain intensity: one for light, two for moderate and two for heavy rains (we do not have violent rains in the studied period). As we can observe in Figure 8, the mean and the range of values of ^{7}Be content are similar for the five constructed categories, although the second category (> 2 to ≤ 5 mm h^{-1}) shows a higher variability than the others.

It is widely recognized that intensity and rain duration are related, when rain duration increases its mean intensity decreases. With this in mind we can evaluate these two variables jointly. The rains can be clustered according to its mean intensity and its duration in short (<5 h) and long rains (>5 h) and the results are shown in Figure 9.

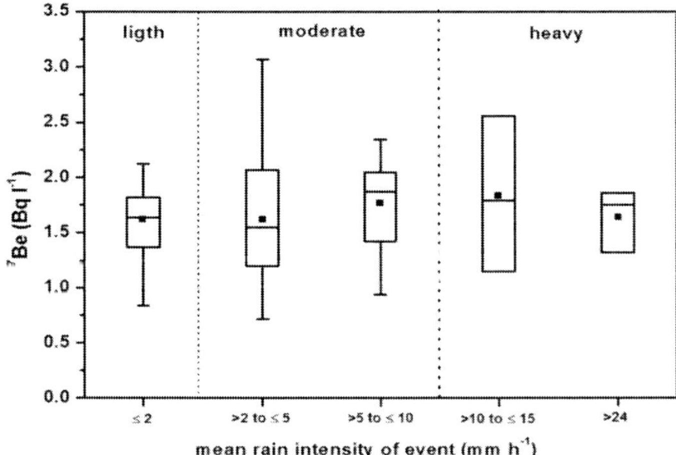

Figure 8. Box chart representation of ^{7}Be activity concentration in rainwater for rainfall events clustered according to mean rain intensity. Boxes are determined by the 25th and 75th percentiles. Whiskers are determined by the 5th and 95th percentiles. The arithmetic means of ^{7}Be activity concentration values are also represented.

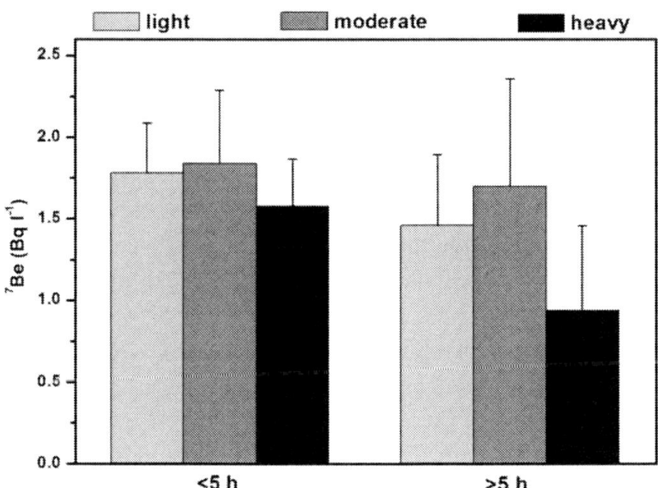

Figure 9. ^{7}Be activity concentration in rain samples for rainfall events of different mean rain intensities and duration.

For rainfalls less than 5 hours long no effect of rainfall intensity was found. Despite no statistical difference was detected for rainfalls longer than 5 hours differences on ^{7}Be content seem to emerge for heavy rainfalls, compared to light and moderate rainfalls. Moreover, if we compare within each category of mean intensity between short and long rain, we see a similar pattern, only heavy rainfalls show a slight difference in ^{7}Be content with higher values in shorter rains.

Ioannidou & Papastefanou (2006) report opposite results, these authors found that short duration and low intensity rainfalls show almost 5 times higher ^{7}Be content than other rains. Again the comparison between these results and ours is difficult because we do not know in detail which one was evaluated by these authors as high and low precipitation intensity.

BERILLYUM-7 WET DEPOSITION

The precipitation magnitude (mm) denotes the height of the water layer which fell on a 1 m^2 ground surface. Hence a 1 mm rainfall indicates 0.001 m^3 (or 1 litter) of rain fallen. Multiplying the ^{7}Be content in rainwater (Bq l^{-1}) by the precipitation magnitude (mm) we obtain the atmospheric deposition expressed in Bq m^{-2}, as follows:

$$Atmospheric\ Deposition\ [Bq\ m^{-2}] = {}^{7}Be\ content\ [Bq\ l^{-1}] \times precipitation\ magnitude\ [mm]$$

As shown in the previous section, researchers find differences and a high variability in measuring ^{7}Be content in rain. But they agree in the general linear pattern when we plot atmospheric deposition versus precipitation magnitude. The Argentinean data also exhibit a linear pattern, as we can observe in Figure 10.

Similar results have been reported for winter season in Japan (Ishiwaka et al., 1995), and for annual periods in Switzerland (Vogler et. al, 1996; Caillet et al., 2001), Greece (Ioannidou & Papastefanou, 2006), Australia (Wallbrink & Murray, 1994). For each study site a clear and characteristic linear relation between thc magnitude of precipitation and the ^{7}Be atmospheric deposition has been founded. Therefore, if the ^{7}Be rainwater content in a study area has been characterized, we can estimate the atmospheric deposition of ^{7}Be.

Table 2. ^{7}Be annual flux and annual precipitations at different locations

Location	annual precipitation (mm)	^{7}Be annual flux (Bq m^{-2} year^{-1})	Reference
	Northern hemisphere		
Bermuda	1700	2850	Turekian et al., 1983
Germany (Heidelberg)	810	1249	Schumann & Stoeppler, 1963
Greece (Thessaloniki)	423	736	Ioannidou & Papastefanou, 2006
India (Bombay)	2277	1267	Lal et al., 1979
Italy (Bologna)	707	1243	Bettoli et al., 1995
Monaco	622	1260	Pham et al., 2013
Netherlands	950	1583	Bleichrodt & Van Abkoude, 1963
Spain (Cantabria)	1268	1670	Rodenas et al., 1997
Spain (Granada)	250	469	González-Gómez et al., 2006
Spain (Huelva)	550	834	Lozano et al., 2011
Spain (Malaga)	550	412	Dueñas et al., 2001
Switzerland (Geneva)	1000	2087	Caillet et al., 2001
Switzerland (Lake Constance)	850	2250	Vogler et al.,1996
Switzerland (Lake Zurich)	1100	2660	Schuler et al., 1991
Switzerland (Rhône River)	1200	2750	Dominik et al., 1987
Syria (Damascus)	153	528	Othman et al., 1998
United Kingdom (Chilton)	822	898	Peirson, 1963
United Kingdom (Milford Haven)	1328	1618	Peirson, 1963
USA (Arkansas)	1070	867	Lee et al., 1985
USA (California)	781	1900	Conaway et al., 2013
USA (College Station - Texas)	1080	2310	Baskaran et al., 1993
USA (Connecticut)	1390	3780	Turekian et al., 1983
USA (Detroit)	739	1900	McNeary & Baskaran, 2003
USA (Florida)	1170	1468	Baskaran & Swarzenski, 2007
USA (Galveston - Texas)	1170	2451	Baskaran et al., 1993
USA (Maryland)	1304	2170	Kim et al., 2000
USA (Massachusetts)	647	2130	Benitez-Nelson & Buesseler, 1999
USA (New Jersey)	787	717	Walton & Fried, 1962
USA (New Hampshire)	645	2767	Benitez-Nelson & Buesseler, 1999
USA (Virginia)	1315	2007	Todd et al., 1989
	Southern hemisphere		
Antarctica	210	700	Nijampurkar & Rao, 1993
Argentina (San Luis)	700	1140	Juri Ayub et al., 2009
Australia (Canberra)	660	1030	Wallbrink & Murray, 1994
Brazil (Rio Janeiro)	865	498	Sanders et al., 2011

Using deposition units (Bq m^{-2}) allow us to evaluate the ^{7}Be annual flux from the atmosphere, and this makes reliable the comparison among different regions, thus avoiding the high variability in ^{7}Be content in rain. Table 2 reports the ^{7}Be annual flux registered in the bibliography, showing firstly that there is lacking information about ^{7}Be for the southern hemisphere relative to the northern hemisphere. On the other hand, from this data, no relationship was found between latitude and annual fallout of ^{7}Be, such as was reported by Wallbrink & Murray (1994). However, only 27% of the variability on ^{7}Be atmospheric deposition may be explained by annual rainfall.

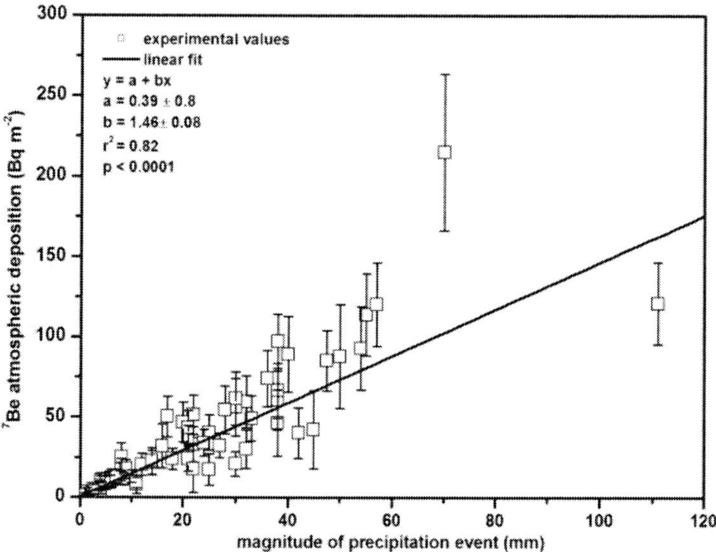

Figure 10. ^{7}Be atmospheric deposition versus magnitude of the rain event. Linear fit was obtained with error as weight. Extracted from Juri Ayub et al., 2012.

As has been reported for many regions of the world when single events are analyzed, the strong correlation between ^{7}Be atmospheric deposition and rainfall magnitude suggests that the global behavior of ^{7}Be could be explained by simple and general laws which take into account rainfall regime and geographic locations (latitude) as main factors. In order to confirm this assertion we need more complete information covering a larger number of environments on the earth.

CONCLUSION

Beryllium-7 is a cosmogenic radionuclide, with relatively short-lived, produced in the stratosphere and the troposphere. It is removed continuously from the atmosphere, mainly by wet deposition and to a lesser extent by dry deposition, as well as radioactive decay. Once ^{7}Be reaches the ground surface is fixed to the fine soil particles. This radionuclide is used as a tracer of atmospheric, geochemical, erosional and sedimentological processes.

With the aim to use ^{7}Be as an environmental tracer, the knowledge of factors that can affect the ^{7}Be income and its total content in soil are needed.

Since wet removal is the main mechanism by which ^{7}Be enters the ecosystems, the measurements of ^{7}Be content in precipitations provide a good tool for understanding these processes.

There is a wide range of values of ^{7}Be content in rainfall, showing the variability of this parameter with the geographical location of the study sites and the season (higher values in spring and summer than in autumn and winter), due to the fluctuations of its atmospheric content. The most important factors that are recognized to affect the ^{7}Be content in rains are: rainfall magnitude, rainfall intensity, elapsed time between rain events and rain duration.

When the ^{7}Be content in rainfalls is analyzed with respect to the effects of rainfall magnitude, elapsed time between rain events and rain duration, two different and conflicting results appear. On one hand the results suggest that the atmosphere could be washed of ^{7}Be when the elapsed time between events is short, or the precipitations magnitude is greater or the rain duration is longer. On the contrary, other studies show no changes in ^{7}Be content with these parameters, suggesting that in some regions the ^{7}Be content in the atmosphere is so great that no washing effect is observed or most likely, that the reload rate of the atmosphere with new ^{7}Be is very short. Another factor that has been suggested to have effect on ^{7}Be content is the rainfall intensity. Again we note that the results are not homogeneous, i.e. different results are obtained for different environments. In the last case the comparison between different studies is difficult.

We need a more complete database with the aim to completely understand the behavior of ^{7}Be content in rains and the variables that have relevant effect on it.

Despite the high variability among different researches, there is a general agreement about the ^{7}Be deposition pattern on the ground due to rainfalls: the atmospheric deposition of ^{7}Be can be reliably fitted to a linear relationship with the magnitude of precipitation events. A clear and characteristic linear relation is found for each study site.

When the annual flux of ^{7}Be from the atmosphere is evaluated no relationship with the latitude is found, but the annual rainfall only explains a small part of the variability on ^{7}Be atmospheric deposition. On the other hand there is lack of information for the southern hemisphere relative to the northern hemisphere.

The strong correlation between ^{7}Be atmospheric deposition and the rainfall magnitude suggests that the global behavior of ^{7}Be could be explained by simple and general laws that take into account rainfall regime and geographic locations (latitude) as main factors. To be sure we need more complete information covering a larger number of environments on the earth.

References

Andrello AC, Appoloni CR. 2010. Measurements of the fallout flux of beryllium-7 and its variability in the soil. *Brazilian Archives of Biology and Technology,* 53 (1): 179 – 184.

Baskaran M, Coleman CH, Santschi PH. 1993. Atmospheric Depositional Fluxes of ^{7}Be and ^{210}Pb at Galveston and College Station, Texas. *Journal of Geophysical Research*, 98 (DII): 20,555-20,571.

Baskaran M, Swarzenski PW. 2007. Seasonal variations on the residence times and partitioning of short-lived radionuclides (^{234}Th, ^{7}Be and ^{210}Pb) and depositional fluxes of ^{7}Be and ^{210}Pb in Tampa Bay, Florida. *Mar. Chem,* 104: 27-42.

Benitez-Nelson CR , Buesseler KO. 1999. Phosphorus-32, phosphorus-37, beryllium-7, and lead-210: Atmospheric fluxes and utility in tracing stratosphere/troposphere Exchange. *Journal of Geophysical Research D: Atmospheres,* 104 (D9): 11745-11754.

Bettoli MG; Cantelli L, Degetto S, Tositti L, Tubertini O, Valcher S. 1995. Preliminary investigations on ^{7}Be as a tracer in the study of environmental processes. *Journal of Radioanalytical and Nuclear Chemistry*, 190 (1): 137-147.

Bhandari, N, Lal D, Rama. 1970. Vertical structure of the troposphere as revealed by radioactive tracer studies, *J. Geophys. Res*, 75 (15): 2974–2980.

Bleichrodt JF, Van Abkoude ER. 1963. On the deposition of cosmic-ray produced beryllium. *J. Geophysical Research*, 68(18): 5283-8.

Brown LI, Stensland GJ, Klein J., Middleton R. 1988. Atmospheric deposition of ^{7}Be and ^{10}Be. *Geochemica et Cosmochimica Acta*, 53: 135-42.

Caillet S, Arpagaus P, Monna F, Dominik J. 2001. Factors controlling ^{7}Be and ^{210}Pb atmospheric deposition as revealed by sampling individual rain events in the region of Geneva, Switzerland. *J. Environ. Radioact*, 53: 241–56.

Cannizzaro F, Greco G, Raneli M, Spitale MC, Tomarchio E. 2004. Concentration measurements of ^{7}Be at ground level air at Palermo, Italy - Comparison with solar activity over a period of 21 years. *Journal of Environmental Radioactivity*, 72 (3): 259 – 271.

Conaway CH, Storlazzi CD, Draut AE, Swarzenski PW. 2013. Short-term variability of ^{7}Be atmospheric deposition and watershed response in a Pacific coastal stream, Monterey Bay, California, USA. *Journal of Environmental Radioactivity*, 120: 94-103.

Dominik J, Schuler C, Santchi PH. 1987. Transport of environmental radionuclides in an alpine watershed. *Earth and Planetary Science Letters*, 84: 165-18.

Dueñas C, Fernández MC, Carretero J, Liger E., Cañete S. 2002. Atmospheric de position of ^{7}Be at a coastal Mediterranean station. *Journal of Geophysical Research,* 106 (D24): 34,059-34,065.

Feely HW, Larsen RJ, Sanderson CG. 1989. Factors that cause seasonal variations in beryllium-7 concentrations in surface air. *J. Environ. Radioact*, 9: 223–49.

Gonzalez-Gomez C, Azahra M, Lopez-Peñalver JJ, Camacho-García A, Bardouni TE, Boukhal H. 2006. Seasonal variability in ^{7}Be depositional fluxes at Granada, Spain. *Applied Radiation and Isotopes*, 64: 228–234.

Ioannidou A, Manolopoulou M, Papastefanou C. 2005. Temporal changes of ^{7}Be and ^{210}Pb concentrations in surface air at temperate latitudes (40° N). *Appl. Radiat. Isot.*, 63: 277–84.

Ioannidou A, Papastefanou C. 2006. Precipitation scavenging of ^{7}Be and ^{137}Cs radionuclides in air. *J. Environ. Radioact.*, 85: 121–36.

Ishikawa Y, Murakami H, Sekine T, Yoshihara K. 1995. Precipitation scavenging studies of radionuclides in air using cosmogenic ^{7}Be. *Journal of Environmental Radioactivity,* 26 (1): 19-36.

Johnson WB, Viezee W. 1981. Stratospheric ozone in the lower troposphere. 1. Presentation and interpretation of aircraft measurements. *Atmospheric Environment* 15 (7): 1309–1323.

Juri Ayub J, Di Gregorio DE, Velasco H, Huck H, Rizzotto M, Lohaiza F. 2009. Short-term seasonal variability in ^{7}Be wet deposition in a semiarid ecosystemof central Argentina. *J. Environ. Radioact.*, 100: 977–81.

Juri Ayub J, Lohaiza F, Velasco H, Rizzotto M, Di Gregorio D, Huck H. 2012. Assessment of ^{7}Be content in precipitation in a South American semi-arid environment. *Science of the Total Environment*, 441: 111 – 116.

Kaste JM, Norton SA, Hess C. 2002. Environmental chemistry of beryllium-7. *Rev Mineral Geochem*, 50: 271–89.

Kim G, Hussain N, Scudlark JR, Church TM. 2000. Factors ifluencing the atmospheric depositional fluxes of stable Pb, ^{210}Pb, and ^{7}Be into Chesapeake Bay. *Journal of Atmospheric Chemistry*, 36: 65–79.

Lal D, Malhotra PK, Peters B. 1958. On the production of radioisotopes in the atmosphere by comic radiation and their application to meteorology. *J. Atmos Sol. Terr. Phys.*, 12: 306–28.

Lal D, Nijampurkar N, Rajagopalan G, Somayajulu BLK. 1979. Annual fallout of 32Si, ^{210}Pb, 22Na, 35S and ^{7}Be in rains in India. *Proceedings of the Indian Academy of Science*, 88: 29-40.

Lee SC, Saleh AI, Banavali AD, Jonouby LJ, Kuroda PK. 1985. Beryllium-7 deposition at Fayetville, Arkansas, and excess Polonium-210 from the 1980 eruption of Mount St Helens. *Geochem. J.,* 19: 317-322.

Lozano RL, San Miguel EG, Bolívar JP, Baskaran M. 2011. Depositional fluxes and concentrations of ^{7}Be and ^{210}Pb in bulk precipitation and aerosols at the interface of Atlantic and Mediterranean coasts in Spain. *J. Geophys. Res.*, 116: D18213.

McNeary D, Baskaran M. 2003. Depositional characteristics of ^{7}Be and ^{210}Pb in southeastern Michigan, *Journal of Geophysical Research*, 108 (D7): 4210.

Murray AS, Olley JM, Wallbrink PJ. 1992. Natural readionuclide behavior in the fluvial environment. *Rad. Prot Dosim*, 45: 285 – 288.

Othman I, Al-Masri MS, Hassan M. 1998. Fallout of ^{7}Be in Damascus City. *J. Radioanal. Nucl. Chem*, 238: 187–199.

Peirson DH. 1963. ^{7}Be in air and rain. *J. Geophysical Research*, 68 (13): 3831-2.

Pham MK, Betti M, Nies H, Povinec PP. 2011. Temporal changes of ^{7}Be, ^{137}Cs and ^{210}Pb activity concentrations in surface air at Monaco and their correlation with meteorological parameters. *Journal of Environmental Radioactivity*, 102: 1045-1054.

Pham MK, Povinec PP, Nies H, Betti M. 2013. Dry and wet deposition of ^{7}Be, ^{210}Pb and ^{137}Cs in Monaco air Turing 1998-2010: Seasonal variations of deposition fluxes. *Journal of Environmental Radioactivity*, 120: 45-57.

Pöschl M, Brunclík T, Hanák J. 2010. Seasonal and inter-annual variation of Beryllium-7 deposition in birch-tree leaves and grass in the northeast upland area of the Czech Republic. *Journal of Environmental Radioactivity*, 101: 744-750.

Ródenas C, Gómez J, Quindós LS, Fernández PL, Soto J. 1997. ^{7}Be concentrations in air, rain water and soil in Cantabria (Spain). *Appl. Radiat. Isot.,* 48(4): 545–548.

Salisbury RT, Cartwright J. 2005. Cosmogenic ^{7}Be deposition in North Wales: ^{7}Be concentrations in sheep faeces in relation to altitude and precipitation. *J. Environ. Radioact.*, 78: 353–61.

Schumann G, Stoeppler M. 1963. Beryllium in the atmosphere. *J. Geophysical Research*, 68(13): 3827-30.

Shuler C, Wielend PH, Santshi PH, Sturn M, Lueck A, Bollhalder S, Beer J, Bonani G, Hofmann HJ, Sutre M, Wolfi W. 1991. A multitracer study of radionuclides in Lake Zurich, Switzerland, 1, Comparison of atmospheric and sedimentary fluxes of ^{7}Be, ^{10}Be, ^{210}Pb and ^{137}Cs. *J. Geophys. Res.* 96: 17051–17065.

Todd JF, Wong GTF, Olsen CR, Larsen IL. 1989. Atmospheric depositional characteristics of beryllium-7 and lead-210 along the southeastern Virginia coast. *J. Geophys. Res.,* 94: 11106 - 11116.

Turekian KK, Benninger LK, Diane EP. 1983. ^{7}Be and ^{210}Pb total deposition fluxes at New Haven Connecticut and at Bermuda. *J. Geophysical Research,* 88: 5411-5.

Vogler S, Jung M, Mangini A. 1996. Scavenging of ^{234}Th and ^{7}Be in Lake Constance. *Limnology and Oceanography*, 4 l(7): 384-1393.

Wallbrink PJ, Murray AS. 1994. Fallout of ^{7}Be in South Eastern Australia. *J. Environ. Radioact.,* 25: 213–28.

Walton A, Fried RE. 1962. The deposition of berillyum-7 and phosphorus-32 in precipitation at north temperate latitudes. *J. Geophys. Res.*, 67: 5335-5340.

INDEX